Professor Weissman's LAUGH with MATH

by

PROFESSOR MARTIN WEISSMAN & KEITH MONSE

LAUGH & LEARN • 5 MEADOWLARK LANE • HACKETTSTOWN, NJ 07840 • (908)813-8299 (718)698-5219

Professor Weissman's LAUGH with MATH
Laugh & Learn / September 1992, September 1995
All rights reserved. Copyright ©1992, 1995 by Professor Martin Weissman and Keith Monse

*No part of this book may be reproduced or transmitted in
any form or by any means, electronic or mechanical, including photocopying,
recording or by any information storage and retrieval system without written
permission from the authors and publisher.*

*ISBN 0-9632003-0-5
Library of Congress Catalog Card Number: 91-78205*

Printed in the United States of America

Panel 1: Good morning everybody... I'm Professor Weissman and welcome to my English Lit class. **GOTCHA!!** I (he,he) get **such** a kick watching everyone's faces momentarily react in horror as you all start fumbling for your schedules. Well...just relax... this **IS** Algebra and you're all in the right place.

Panel 2: **HEY!** Where are you going?

Panel 3: Gotta go man...I really do got English Lit now!

Panel 4: Geez...that's never backfired on me like **that** before! But anyway...I see a lot of interesting faces out there, as well as a couple of noteworthy elbows, ankles and earlobes, but we'll stick with the faces. Any of you want to stand up and tell us a little something about yourselves? How about you?

Panel 5: Well...Professor...I took Algebra in High School. If you asked me what I got out of that class, I'd **probably** reply "about 40 winks." (yuk yuk)

Panel 6: As for me...it's really peculiar but some powerful force keeps telling me that I should brush up on my basic Algebra skills before continuing my education.

Panel 7: That's very commendable. Just exactly what **is** that powerful force?

Panel 8: The college dean! He's 6'5", 260 pounds, and built like a brick...

—1—

er...I get the picture. Welcome to the class. What about you sir?
I never got da chance to graduate but I plan on taking the GED and I need to learn dis here Algebro stuff.
That's Algebra!
Hey look everybody...I'm loining already! You're okay dere professor.
Thanks...I try. Anybody else?
Well...I'm retired going on two years now. I really need to get my brain working again. I tried watching "Geraldo" but that just did more damage, So I figured an Algebra class would be the perfect outlet to reenergize my thinking skills and give me the same satisfaction that logic and problem solving gave me during my career.
MATH
Problem solving? You must have been some kind of big business tycoon or computer troubleshooter...right?
Actually...I sold knishes, but I never could decide which mustard to keep in that big dispenser jar...the yellow one or that spicy brown kind!
HA
HO
HA
HA
HE
OHHHH...KAY! I think it's time we move on.

-3-

Well besides that...if we extend the line to the left we include NEGATIVES.

THE INTEGERS

The positive and negative whole numbers as well as the zero make up what we call **"THE INTEGERS."**

Can anybody give us an example where positives & negatives occur in the real world?

How about a weather thermometer? Right?

Sometimes you see them on signs. On my last vacation I saw a sign that read "1200 feet below sea level." **That's -1200!**

Good examples. You might not have noticed it yet but I'm on a diet and have lost 5 pounds so far! **That's -5!**

YOU'RE RIGHT! We haven't noticed it yet!

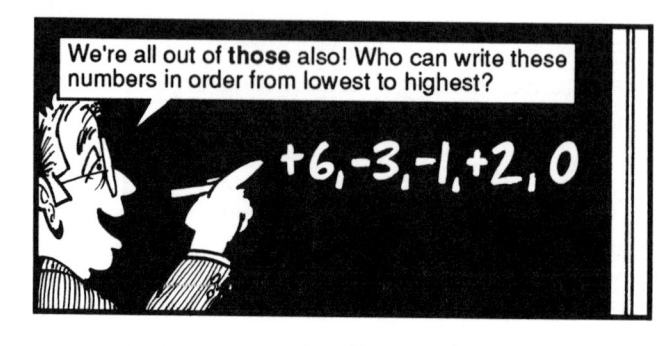

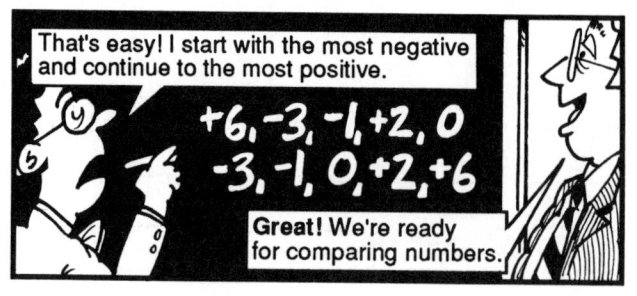

What about the stock market reports Professor? According to today's newspaper, Magnum Corporation is +4, but General Industrial is -35!

DID YOU HAVE TO REMIND ME OF THAT?

O.K. class...let's talk about ordering now!

ALRIGHT! I'll have a pastrami on rye, hold the mayo with a side order of fries...

Sorry my hungry friend but we're all out of that. Besides, I was referring to ordering the **NUMBERS!**

RATS!

We're all out of **those** also! Who can write these numbers in order from lowest to highest?

+6, -3, -1, +2, 0

That's easy! I start with the most negative and continue to the most positive.

+6, -3, -1, +2, 0
-3, -1, 0, +2, +6

Great! We're ready for comparing numbers.

Panel 1: Just remember...if we take any two numbers, the greater one would be the number farther to the right. "**>**" means "**is greater than**."

+5 > +2

As you can see, +5 is **greater** than +2 because +5 is more to the **RIGHT**.

Panel 2: **HEY!** Are you making some kind of political commentary?

Panel 3: Calm down...here watch...check it out.

Panel 4: +5 > +2
+2   +5

What symbol would you use here? How about you young man...wanna give it a shot?

Sure..I remember that!

Panel 5: +5 > +2
+2 < +5

We reverse the symbol because +2 **"is less than"** +5.

Panel 6: I always confuse the < and > symbols. Then again, I can **never** remember which Gabor sister starred on "Green Acres." ZsaZsa or Eva?

It was Eva! ZsaZsa's never starred in **anything**! But back to those symbols. Here's how I remember them. > is on the right of the number line and numbers on the right are bigger. < is on the left of the number line and numbers on the left are smaller.

Panel 7: Not bad...I remember < because it looks like an **L**. See...

<LESS

Panel 1: Hey...how about another problem Professor?

Panel 2: Sure...which symbol, < or >, would you use between **these** numbers?
-4 -2

Panel 3: -4 is **greater** than -2.
No...wait! -4 is **less** than -2!
One of you is right. Young lady, come up to the board and fill in the proper symbol.

Panel 4: Less than because -4 is to the **left** of -2!
-4 < -2

Panel 5: Yo...wait a second! Isn't 4 **bigger** than 2?

Panel 6: Yes...but remember to watch those **signs**. -4 is less than -2!
-4 < -2

Panel 7: Our friend here ignored the signs. That's interesting because sometimes we **aren't** concerned about the signs.

Sounds like my last road test Professor Weissman. (he he)

When you're **not** concerned about the signs...isn't that called **absolute value**?

Absitively! Here, check out my handy dandy diagram.

—6—

Now...if you're low on gas and want to get to town fast, which way would you go?
WEST
GAS 8 MILES
EAST
GAS 3 MILES
I don't know about anybody else, but **I'd** grab my siphon and find the nearest parked car.
You'd go to the right. It's only 3 miles. To the left it's 8 miles.
Of course...-8 is less than +3 but you don't care about the direction or the sign...just the magnitude or the **absolute value.**
-WEST 8 +EAST 3
-8 < +3
|-8| > |+3|
8 > 3
So those vertical lines mean **absolute value?** Sort of "drop the sign?"
As my wife said when she pulled the car out of the garage, "I **think** I'm making a dent!"
As we see...numbers in Algebra have **two** parts. A sign and a magnitude or "absolute value."
|-8| > |+3|
8 > 3
SIGN
MAGNITUDE
The sign shows direction and the magnitude tells us **"how much."** Has anybody here used signed numbers lately in their everyday lives?
My wife Mildred wrote a check yesterday for 350 dollars...look at this entry.

CHECK DATE DESCRIPTION + PAYMENT − WITHDRAWAL $BALANCE
28 2/15 TRANSFER $300
29 2/28 ELECTRIC COMPANY $50
30 2/28 DEPOSIT $50
TRANSFER $200
PHONE COMPANY $200
3/5 FUR SALON $350
33 3/10
Good example. The "-" is the direction. It means she withdrew money and the 350 is the magnitude or how much!
There are also some "+" entries. Those are the deposits...right?
Right! But, I notice our friend hasn't done the balance column for some time. Let me show you how to combine signed numbers.
First we combine all the deposits which are positives...then combine the negatives...
+300 -50
+50 -200
+200 -350
+550 -600
It's easy to combine the numbers when all the signs are the same. This is an example of CASE 1!
CASE 1
LIKE SIGNS
Add The Absolutes
Keep The Common Sign
+550 -600
As you notice... the two results have opposite signs. To combine them is a scenario for CASE 2!
CASE 2
UNLIKE SIGNS
Subtract The Absolutes
Use The Sign Of The Larger
So we combine the resulting numbers as follows.
|+550|-|-600|
600-550=50
-50
Subtracting the absolute numbers we get a result of 50. Then we must remember to use the sign of the number with the larger absolute which was a minus. This will give us a final answer of -50. Uh Oh...looks like you're a little short in your checking account. I think you better make a quick deposit.

DANG YOU MILDRED!

Well...that's the way the check bounces.

Let's continue...

As I mentioned before...I'm on a diet. Can anybody tell me how I'm doing so far?

MY DIET
WEEK 1 −2 lost
WEEK 2 −5 lost
WEEK 3 +1 gained
WEEK 4 −2 lost
WEEK 5 +3 gained

I'll give it a try. That's as easy as pie!

DID YOU HAVE TO SAY PIE?

Let's see now...first we combine **all** the negatives... then **all** the positives. Finally, we combine the resulting numbers, the −9 and the +4, by **subtracting** their absolute values and using the sign of the larger, giving us −5.

$$\begin{array}{cc} -2 & +1 \\ -5 & +3 \\ -2 & +4 \\ \hline -9 & \end{array} \quad \begin{array}{c} 9-4 \\ \hline 5 \\ -5 \end{array}$$

Hey Professor! You got any diet tips for **me**?

Yeah...you can **start** with a shave and a haircut!

Settle down everybody...let's backtrack a little bit!

$$|+7| = 7$$
$$|-7| = 7$$
$$|0| = 0$$

Remember absolute values?

−9−

That's when we drop the sign... right?
Right...but let's be more precise. The absolute of a positive is...
POSITIVE!
...and the absolute of a negative is...
ALSO POSITIVE!
...and the absolute of zero is...
ZERO?
|+7|=7
|-7|=7
|0|=0
That's right...the absolute value is ALWAYS POSITIVE or ZERO! However, the procedure is a little different when you have more than one number between the absolutes... watch!
|-5+3|
|-5+3|
|-2|
2
First we simplify the numbers between the absolutes. -5 + 3 gives us -2. Now, the absolute value of -2 is 2, which is our answer.
Professor...I'm confused!
Are you positive?
I don't know if I'm positive or negative! I'm just confused!
He's always in the dark Professor.
IN THE DARK? Say... that gives me an idea.

CLICK

OK folks...it's time to learn about **subtraction**

IN THE DARK?

HEY! Who turned out the lights?

Whoa! Whose hand is that?

I'm scared!

HEY... somebody took my good pen!

OK...lights on! Now... if you'll excuse my graffitti, I'd like to make a point.

Subtraction acts like this switch. Think of the switch as a **subtraction symbol**.

SUBTRACTION SWITCH
+ ON [ ] OFF −
"Think Opposite!"

Look at the light. If it's **on** (or positive) you flip the switch (subtract) and the light is **off** (or negative).

And if the light is off or **negative** and you flip the switch again or subtract, then the light is on or **positive**.

Never mind the light switch...I think the **Professor** has flipped.

I heard that wiseguy... here let me make it real clear for you!

+10−(−8)

See this subtraction sign. Picture it as that light switch because it will **change** the sign that follows.

OBSERVE...

+10−(−8)
+10 +8

Now we combine.

+10−(−8)
+10 +8
+10 +8 = +18

The answer is +18!

Hey everybody, check it out!

To Subtract
A poem by Lenny Krinsky

Change The Sign, Then Combine

The End

−11−

Panel 1:
Right...he's a poet and doesn't even know it (he he).

I GOT ONE PROFESSOR! 'Dere once was a man from Nantucket...

Panel 2:
That's **quite** enough Longfellow. Anyhow, did you see what happened to the parentheses and the sign outside.

Panel 3:
$-(-8)$
$+10 + 8$
$+10 + 8 = +18$

As we just saw, a negative outside of the parentheses **changes** the sign that follows. The -8 becomes a +8 and we **remove** the negative sign outside **and** the parentheses as well.

Panel 4:
However, a positive outside of the parentheses DOES NOT change the sign that follows.

$+(-15)$
-15

So, in this problem, the -15 **remains** -15 but, we here also, **remove** the outside positive sign **and** the parentheses.

Panel 5:
$-(9-2)$

Now...in this example, the first negative sign will change **BOTH** signs that follow.

Panel 6:
The 9 has no sign. That means it's **positive** so it becomes -9 and the -2 now changes to +2.

$-(9-2)$
$-9+2$

Again, we **remove** the outside negative sign **as well as** the parentheses.

Panel 7:
Can't we finish the problem Professor?

Pssssst... keep it down ALGEBRA BREATH!

Panel 8:
Sure...just combine, remember? We come up with -7.

$-(9-2)$
$-9+2$
-7

Now...let's get a little bit **more** complicated!

—12—

-(6-9)+(-8)-(-7+11)-(-9)
YIKES!
That better be somebody's phone number Professor.
I feel a migraine coming on.
HEY! Isn't that the combination to my gym locker?
NEVER MIND THAT! I think that's my wife's measurements!
Here...let me give you a tip!
Blue Bell in the fifth?
RACING FORM
Not exactly...first separate the problem into smaller parts or "TERMS."
-(6-9) | +(-8) | -(-7+11) | -(-9)
See...it has 4 terms!
-(6-9) | +(-8) | -(-7+11) | -(-9)
-6+9 | -8 | +7-11 | +9
+3 | -8 | -4 | +9
Now...we simplify each term, finally we'll combine the four results.
-(6-9) | +(-8) | -(-7+11) | -(-9)
-6+9 | -8 | +7-11 | +9
+3 | -8 | -4 | +9
+3+9=+12
-8-4=-12
0
Remember combining? And now we have our answer.
HEY PROFESSOR! You must be real disappointed by that answer!
Why do you say that?
The answer was zero... all that work for NOTHING!

Hey Prof... is the answer **+0** or **-0?**
Just zero without a sign. Zero is the **only** interger with no sign.
It's time for those old **VISUAL AIDS!**
That's sticking to your diet Professor!
Never mind that! Just solve this problem.
3a+2a=
That's easy...**5a².**
Are you **sure?** Here...watch!
WHOA! Here are 3 apples or as we'll call them...**3a.**
Now...if I pick up 2 **more** apples or another **2a...**
BONK
...I'll ask again! What's **3a + 2a?** And **please,** don't anybody say "It's Professor Weissman about to make applesauce!"
I got it... the answer is 5 apples, er...I mean 5a.
Right...just 5a. **NOT SQUARED!** Besides, apples are round anyway (he he).
3a+2a=
So when do we get **a²** Professor?
When we multiply! But we'll do that later...right now we're combining. We say that 3a and 2a are **LIKE TERMS!** They have the same letters. Just combine the numbers and keep the letter.
COEFFICIENTS
3a+2a=5a
LITERAL FACTORS
The numbers are called **"COEFFICIENTS"** and the letters are called the **"LITERAL FACTORS!"**

How about some more brain exercise? Take a look at **these** examples...
#1 5b3+3b3=8b3
#2 2x+7x+x=10x
#3 6ab+7ab=13ab
#4 9xy2+5xy2=14xy2
...and like the great guy I am, I'll even provide all the answers.
Hey Professor... where **were** you when I was taking the SAT?
Or when my father caught me sneaking home 3 o'clock in the morning?
I don't understand something Professor. Why is the answer to the second problem **10x? I get 9x!**
ZZZZZ
9x
That's a common mistake. When a term has no number...put in a 1.
So the example will read **2x+7x+1x=10x**
What about the first example? Shouldn't the answer be **8b6?**
Again...that's what we do when we multiply. Remember, we're combining now!
Now...back to my fruit again. I've got **3 apples** and 2 **bananas.** What do I get when I combine them?
3a+2b=
5ab?
I know how we can combine those apples and bananas Professor!
I won't even **ask** what you're doing with a blender. The two terms, **3a** and **2b** are **UNLIKE!** Unlike terms **can't** be combined. You don't have 5 apples and you don't have 5 bananas and you certainly don't have 5 applebananas! Ain't no such kind of fruit!
SO WHAT'S THE ANSWER?
There **is** no answer! You can't add apples and bananas... that is unless you want to make fruit salad.

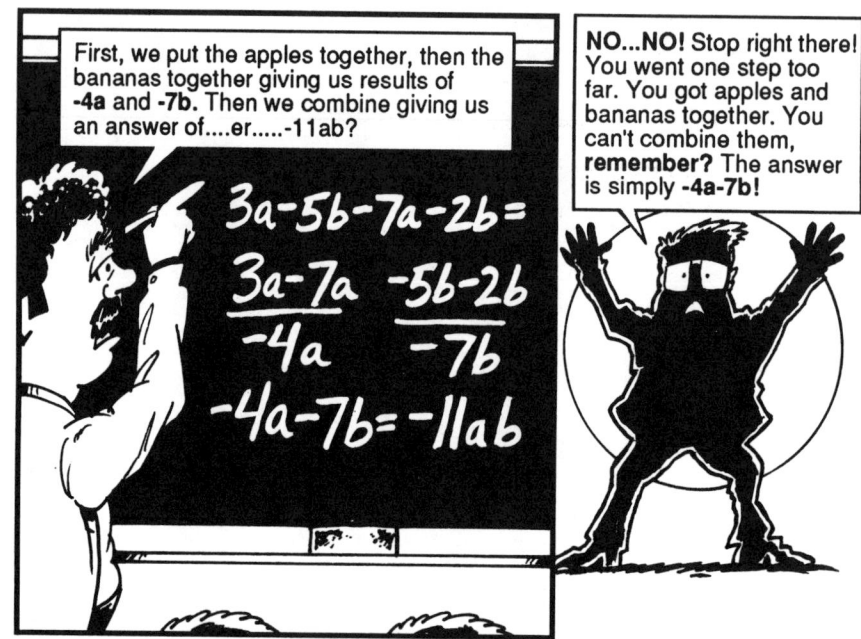

Panel 1:
Let's stay in our fruit mode. Anybody wanna attempt **this** one?

$3a - 5b - 7a - 2b =$

I used to work in a produce market. I guess that qualifies me.

Panel 2:
First, we put the apples together, then the bananas together giving us results of **-4a** and **-7b**. Then we combine giving us an answer of....er.....-11ab?

$3a - 5b - 7a - 2b =$
$\underline{3a - 7a} \quad \underline{-5b - 2b}$
$-4a \quad\quad -7b$
$-4a - 7b = -11ab$

Panel 3:
NO...NO! Stop right there! You went one step too far. You got apples and bananas together. You can't combine them, **remember?** The answer is simply **-4a-7b**!

Panel 4:
$3ab + 7ba =$

One final problem!

NO ANSWER! Can't be done! Right? Those are **UNLIKE TERMS**.

Wrong! The answer is 10ab.

HUH?

Panel 5:
Allow me to elucidate. You see, **ab** actually means **a times b**. It's the same as **ba** because when you multiply the order is not important. I can change the **7ba** to **7ab**. So now both terms look alike and we can combine.

$3ab + 7ba = 10ab$
$3ab + 7ab = 10ab$
$10ab$

Panel 6:
I guess that's what they mean by being fruitful and multiplying. So, Professor, wouldn't **10ba** also be correct?

Panel 7:
$3ab + 7ba = 10ba$
$3ba + 7ba = 10ba$
$10ba$

Yes! It's the same either way! See...

Panel 8:
Well class...I think that's enough for today. We've covered a lot of ground and gotten a strong foothold in Algebra basics. On the following pages are some sample problems for you to try. And **remember**, don't look at the answers first. Professor Weissman will be watching. I'll see you soon for our **next** lesson.

PROBLEM SET #1

See Answer Section in Back

1. In Algebra we use numbers from Arithmetic. We put a plus sign before them and they are now called .... numbers.

2. When we extend the number line from 0 to the left we now include the .... numbers.

3. The positive and negative whole numbers and zero are called the ....

4. The symbol < looks like the letter L and means " is .... than. "

5. The symbol for " is greater than " is ....

6. On the number line, the larger of 2 numbers is more to the ....

7. On the number line the smaller of 2 numbers is more to the ....

REPLACE THE ? IN EACH OF THE FOLLOWING INEQUALITIES WITH THE CORRECT SYMBOL: > or <

8. +9 ? +5

9. +6 ? 0

10. 0 ? -4

11. +7 ? -10

12. -3 ? 0

13. -4 ? -1

14. Two vertical lines that surround a number indicate ....

15. The absolute value of a number is always .... or ....

EVALUATE EACH OF THE FOLLOWING ABSOLUTES:

16. | -9 |

17. | +9 |

18. | 0 |

19. Numbers in Algebra have 2 parts: a .... and a ....

20. If the sign of a number is missing, it is understood to be ....

21. To combine numbers with like signs, ... the .... and use the common sign.

USE THE ABOVE RULE (PROBLEM 21. TO COMBINE IN EACH OF THE FOLLOWING PROBLEMS:

22.   +9 +12

23.   -7 -10

24.   -1 -2 -3

25.   -1 -2 -3 -4

26. To combine numbers with unlike signs, ... the absolute values and use the sign of the number with the .... absolute value.

USE THE ABOVE RULE (PROBLEM 26. TO COMBINE IN EACH OF THE FOLLOWING PROBLEMS:

27.   -12 +8

28.   -9 +1

29.   8 -15

30.   -1 +2 -3 +4 -5

31. 1 -2 +3 -4 +5

32. -10 +10

33. -10.8 +7.2

34. -2/3 +7/8

35. When adding, the opposite of -10 is ....

36. The sum of a number and its opposite is ......

37. The number 0 (zero) has no ....

38. A negative sign that is on the left of parentheses, tells you to remove the parentheses and .... the signs of the numbers inside parentheses.

SIMPLIFY EACH OF THE FOLLOWING:

39. - (-7)

40. - (+8)

41. - (-7 +8)

42. - (-4 -6)

43. - (+3 +10)

44. - (-1 +2 -3)

45. When a positive sign precedes parentheses, remove the parentheses but DO NOT .... the signs of the numbers inside the parentheses.

SIMPLIFY EACH OF THE FOLLOWING:

46. + (-7)

47. + (-3)

48. + (-2 -5)

49. + (-6 +9)

50. + (-1 +2 -3)

51. When simplifying an expression it is best to separate the problem into ....

SPLIT EACH PROBLEM INTO TERMS AND SIMPLIFY.

52. -6 -(+7) +(-8)

53. 10 - (-4) - (-7)

54. 40 - (8-10) +(-6)

In statistics, to find the RANGE of a set of numbers you subtract the smallest number from the largest.

55. If the high temperature one day was 88 degrees and the low was 40 degrees, what was the temperature range?

56. Find the range if the high was 33 degrees and the low was -5 .

SIMPLIFY EACH OF THE FOLLOWING:

57. 2/7 - (-3/7)

58. 2/5 - (+3/5)

59. -7.9 - (-10)

60. 20 - (-1.6)

61. 7/8 - (-3/16)

-18-

62. 0.999 - (1)

USE SIGNED NUMBERS TO SOLVE EACH OF THE FOLLOWING PROBLEMS.

63. Professor Weissman has $46 in his checking account. He writes a check for $16, deposits, $5 and writes a check for $20. What is his balance ?

64. A helicopter is 850 feet above sea level and a deep sea diver directly under it is 30 feet below sea level. What's the distance between them?

65. Mt. Whitney in central California (14,495 ft) is the highest point in the contiguous 48 states. Death Valley, in southeast California (282 feet below sea level) is the lowest point in the nation. Use signed numbers to determine how much higher than Death Valley is Mt. Whitney?

66. On a hot day last summer the temperature dropped from 88 degrees to 59 degrees in 2 hours. How many degrees did it drop?

67. On a cold winter day the temperature rose from -12 degrees to +6 degrees. How many degrees did the temperature rise ?

68. The Dead Sea is 1290 feet below sea level and Death Valley is 282 feet below sea level. How much higher is Death Valley than the Dead Sea ?

69. LIKE TERMS terms are terms which have the same .... raised to the same ....

IN THE NEXT 3 PROBLEMS LOCATE THE LIKE TERMS:

70. $6xy$ $-8x$ $7y$ $6x^2$ $-5xy$

71. $6ab$ $-8b^2a$ $-ab$ $4ba$

72. $5m^2n$ $-2mn^2$ $3nm^2$

73. In the term $-3ab$, the -3 is called the ....

74. In the term $-3ab$ the literal factors are ....

75. If you don't see a numerical coefficient you can assume that it is a ....

IN EACH OF THE FOLLOWING PROBLEMS WRITE NUMERICAL COEFFICIENT.

76. $-8mn$

77. b^2

78. $-k$

79. To combine LIKE TERMS combine the .... and keep the common ....

80. .... terms are terms with different literal factors.

81. UNLIKE TERMS cannot be .....

82. Since factors are multiplied the order of the factors is not important. Another way to write $7xy$ is ......

83. In $-xy$, the numerical coefficient is ....

84. In $-xy$, the x and y are called ....

85. When you combine like terms you do not add the exponents. The sum has the same .... as each like term.

IN EACH OF THE FOLLOWING PROBLEMS COMBINE LIKE TERMS:

86. 7x −18x

87. −6abc −4abc

88. xy −8xy

89. $5a^2 - 2a^2$

90. 2a −5b −7a −b

91. 5mn −8nm

92. $a^2 + a^2$

93. On a test a student wrote
2a + 3b = 5ab
What was his mistake?

94. The same student wrote
$5b^3 + 7b^3 = 12b^6$
What did he do wrong?

95. $5m^2 + 5m^3 = ?$

96. | −7 +2 | = ?

97. |−7| + |+2| = ?

98. − | −7 | = ?

99. − | +7 | = ?

100. − | −7 +2 | = ?

101. −3 −7 −2 = ?

102. −3 +5 −7 +1 −2 +9 = ?

103. 8 −3 −2 +1 −4 = ?

104. 5d +3d = ?

105. 5m +7m +m = ?

106. −5m −7m −m = ?

107. 7n −10n = ?

108. 3m +2n +7m +n = ?

109. 3m −2n −7m −n = ?

110. 5x + 2y = ?

111. $3c^2 + 7c^2 = ?$

112. 4mn +7mn = ?

113. 20ab − ab = ?

114. 5cd + 6dc = ?

115. 3a − 5b − 7b −8a +5ab = ?

116. $7xy^2 - 10xy^2 = ?$

117. $4ab^2 + 6a^2b = ?$

118. −7am +7am = ?

119. 5x −7y +2y −9x +5y = ?

120. 5 − 3x −9 −2x = ?

121. 10m −3 −2 −10m +5 = ?

122. − (−5) = ?

123. − (+6) = ?

124. − (−5 +6) = ?

125. 10 − (−3) = ?

126. + (−4) = ?

−20−

127. $+(+8) = ?$

128. $+(-6+5) = ?$

129. $12-(3-7) = ?$

130. $-(-4)+(-3)-(5) = ?$

131. $40-(5-10)-(-1) = ?$

132. $-(-5a) = ?$

133. $-(2b-1) = ?$

134. $-(-3e-7e) = ?$

135. $2x-(-5x) = ?$

136. $2a-7-(5+6a) = ?$

137. $3m-7+(2m+1) = ?$

138. $(2a-4b) = ?$

139. $(2a-4b)+(6a-7b) = ?$

140. $(2a-4b)-(6a-7b) = ?$

141. Subtract $-3x$ from $7x$

142. Subtract $2a+5b$ from $7a-2b$

143. Find the sum of $-3x$ and $-2x$

144. Find the diffrence of $-3x$ and $-2x$

145. $2.3x+5x+0.16x = ?$

146. $40x-3.4x = ?$

147. $23y-1.71y = ?$

148. $(2/3)x + (4/5)x = ?$

149. $(2\ 1/2)y + (3\ 1/8)y = ?$

150. $(3/4)a + 2.5a = ?$

151. $3p-7p = ?$

152. $7p-3p = ?$

PUT THE NUMBERS IN ORDER FROM SMALLEST TO LARGEST. SIMPLIFY EACH NUMBER FIRST IF NECESSARY.

153. $-3, +2, -7, +5, 0$

154. $|-5|, -(-9), |7-11|$

155. $-13/2, 0, 7.01, -6.1$

SIMPLIFY EACH OF THE FOLLOWING PROBLEMS:

156. $(-5)+(-7)$

157. $|-5|+|-7|$

158. $(-4)-(-3)$

159. $|-4|-|-3|$

160. $-(-[-3])$

161. $-(-[4-9])$

162. Find the sum of $3a+8b-5c$ and $-7a+b+2c$

163. Find the sum of $3x-2y$ and $6x-7y$ and $-4x-y$

164. Add $3-7x-x^2$ and $5x^2-3x-3$

165. How much less than 25 is 12?

-21-

166. How much less than 5x + 4y is 2x + y ?

167. What must be added to +10 to get 0?

168. What must be added to 4x - 7y to get 0 ?

169. What must be added to 20 to get 50 ?

170. What must be added to -14 to get +10 ?

171. What must be added to 8x to get -7x ?

172. How much is 30 decreased by 17 ?

173. How much is -40 decreased by -15 ?

174. How much is 7a decreased by -7a ?

175. How much is 3x -5y decreased by 7x +2y ?

176. How much is 2a -5b increased by -7a -b ?

177. If | X | = 5, what are possible values for x ?

178. If | X - 7 | = 0, what's x equal to?

IS THE GIVEN VALUE FOR X A SOLUTION TO THE INEQUALITY ?

179. | X | < 9 ;   X = - 8

180. | X | > 4 ;   X = - 7

181. | X - 1 | < 6 ; X = - 5

182. | 2 - X | > 0 ;   X = -3

183. The length of a rectangle is 5 and its width is 7. What is the perimeter ?

184. The length of a rectangle is x and its width is 2x . What's the perimeter?

185. The length of a rectangle is x and its width is y. What is the perimeter ?

186. The 3 sides of a triangle are 2x - 5, x and x - 7. What is the perimeter ?

187. A student has 17 nickels. What are they worth (in cents) ?

188. A student has x nickels. What are they worth in cents ?

189. A student has 3x nickels. What is their value in cents?

190. What is the value of 7x dimes ?

191. Joe is 22 years old and Mary is 17. What is the sum of their ages ?

192. Moe is x years old and Susan is x-3. What is the sum of their ages ?

193. Professor Weissman is 49 years old. How old was he 29 years ago when he started teaching ?

194. Professor Weissman's wife is x years old. How old was she 29 years ago ?

DOES THE GIVEN VALUE OF X SATISFY THE INEQUALITY ?

195.   X < 5 ;   X = 7

196.   X > -7 ;   X = -4

197. X < 0 ;   X = -10

198. 0 > X ;   X = -3

199. -7 < X ;   X = -3

THE INEQUALITY 5 < 8 (5 is less than 8) CAN BE REWRITTEN 8 > 5 (8 is greater than 5). REWRITE EACH OF THE FOLLOWING INEQUALITIES.

200. -7 < 0

201. -10 < 7

202. 0 > X

203. -5 < X

THE INEQUALITY 3 < 4 < 5 CAN BE READ " 4 is between 3 and 5. "

More generally, when a number X is between 2 numbers, say a and b we can show this with an inequality:

   a < X < b

FOR EACH OF THE FOLLOWING INEQUALITIES REPLACE X WITH THE GIVEN VALUE THEN STATE WHETHER THAT VALUE OF X SATISFIES THE INEQUALITY ?

204. 5 < X < 12   X = 6

205. -7 < X < 0   X = -3

206. -5 < X < -1   X = -8

207. -3 < X - 6 < +3   X = -2

208. -8 < X + 5 < 0   X = -3

THE SYMBOL $\leq$ MEANS "less than or equal to." $\geq$ MEANS "greater than or equal to." The line under each inequality is the equal sign.

$5 \leq 10$ is true because 5 is less than 10.

$7 \leq 7$ is true because 7 equals 7.

FOR EACH OF THE FOLLOWING INEQUALITIES REPLACE X WITH THE GIVEN VALUE THEN DECIDE WHETHER THAT VALUE SATISFIES THE INEQUALITY ?

209. X - 7 $\geq$ -14   X = -7

210. 8 - X $\leq$ 0   X = -8

211. X + 4 $\geq$ -2   X = -6

212. -7 $\leq$ X < +7   X = -7

213. -7 $\leq$ X < +7   X = +7

Panel 1:

"Welcome back! How'd we all do with the sample problems? I hope none of our readers peeked at the answers. And if you **did**...in the words of the esteemed 20th Century philosopher Gomer Pyle, 'Shame...shame...shame!'"

"As you recall, the last thing we did was to combine **LIKE TERMS**."

"I remember. Only **like terms** can be combined, right?"

"Let's try a refresher!"

"ICE COLD LEMONADE!"

Panel 2:

"Not exactly...combine these four terms! How about it young lady?"

$3x - 5y - 10x - 3y =$

Panel 3:

"First...I'll rearrange the terms."

$$3x - 5y - 10x - 3y =$$
$$3x - 10x - 5y - 3y$$
$$-7x - 8y$$

"Now...I'll combine each group and wind up with an answer of **-7x-8y**."

Panel 4:

"Hey Professor! I got an answer of **-8y-7x**."

Panel 5:

"That's also OK. The terms can be rearranged if you choose. Either way we have to stop because the terms **-8y** and **-7x** are **UNLIKE**! Remember?"

Panel 6:

"Let's move on! Now we're ready to continue with multiplication and division. But before that, a little **nomenclature** is in order."

$8 \times 7 = 56$

"Who's Norman Clature?"

"NOMEN! NOMENCLATURE!"

"WHAT'S THAT?"

—24—

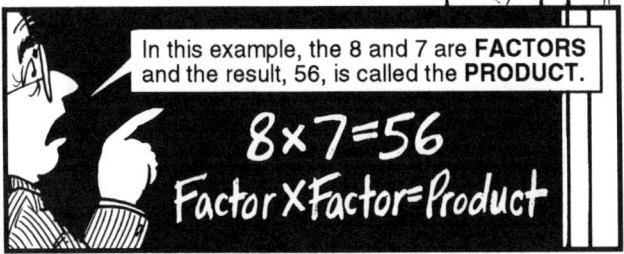

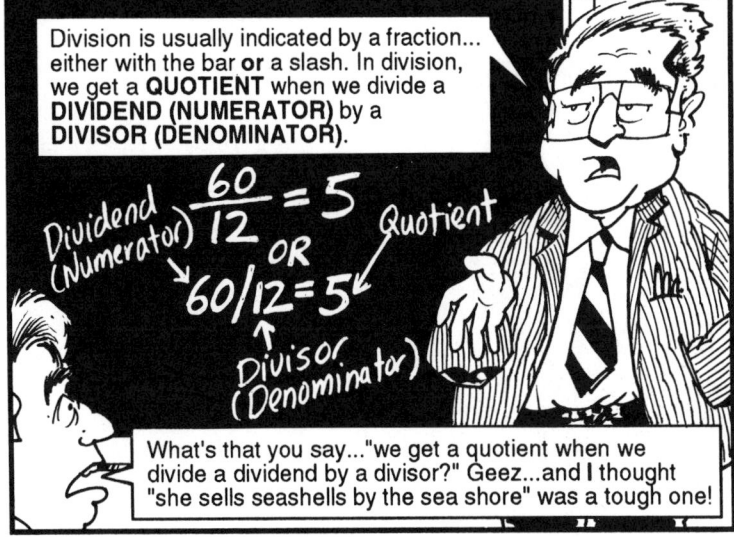

Panel 1:
- "Doesn't **anyone** have a dictionary?"
- "Dictionary? I thought this was math!"
- "Right here Professor Weissman! *'A set of names or terms as those used in a particular science or art.'*"

Panel 2:
- "**Correctamundo!** This is some of the **'nomenclature'** we use in multiplication and division."

MULTIPLICATION	DIVISION
Factors	Dividend (Numerator)
Product	Divisor (Denominator)
	Quotient

Panel 3:
- "In this example, the 8 and 7 are **FACTORS** and the result, 56, is called the **PRODUCT**."

$$8 \times 7 = 56$$
$$\text{Factor} \times \text{Factor} = \text{Product}$$

Panel 4:
- "A little trick is to remember **'FACTORIES MAKE PRODUCTS.'**"

Panel 5:
- "Division is usually indicated by a fraction... either with the bar **or** a slash. In division, we get a **QUOTIENT** when we divide a **DIVIDEND (NUMERATOR)** by a **DIVISOR (DENOMINATOR)**."

$$\text{Dividend (Numerator)} \rightarrow \frac{60}{12} = 5 \quad \text{OR} \quad 60/12 = 5 \leftarrow \text{Quotient}$$
$$\text{Divisor (Denominator)}$$

- "What's that you say...'we get a quotient when we divide a dividend by a divisor?' Geez...and **I** thought 'she sells seashells by the sea shore' was a tough one!"

Panel 6:
- "I wouldn't even **try** to say that! My **uppers** might fall out!"
- HA HO HO HA HA $8 \times 7 = 56$
- "We'll cross that bridge when we come to it. **Get it? Bridge?** (Har Har) Sometimes I just crack myself up! Getting back to our lesson, in Algebra we could have a **real** problem with the **X** when we write **8×7=56**. Can anybody explain **why**?"

Panel 7:
- "Because we use the **X** in equations?"

Panel 8:
- "**YES!** In Algebra we use letters... a, b, c,... x, y, z"

Panel 9:
- "Oh great...now I'm back in nursery school! What's next Professor...**finger painting** or **nappy time**?"

—25—

Allow me to explain. Since we might use **X** instead of a number, we **can't** use the letter **X** when we multiply?

In Algebra, we use the parentheses for multiplication. See!

$$8(7) = 56$$
$$(8)7 = 56$$
$$(8)(7) = 56$$

FACTORS PRODUCT

Occasionally, we can even use a dot to represent multiplication.

$$8 \cdot 7 = 56$$

Just make sure the dot is raised a bit so it won't be confused with a decimal point.

And sometimes, we **don't** need any symbols. For instance...

$$4(b) = 4b$$
$$4 \cdot b = 4b$$

Let's talk in a little bit **more** detail about multiplication and division. We multiply and divide in Algebra like we do in arithmetic. We **just** need some rules for determining the sign of the answer.

As we all know, "Two wrongs **don't** make a right" but..."Two **negatives** make a **positive**...when you multiply **or** divide."

For example...in multiplication... when **BOTH** factors are negative...

$$(-3)(-5) = +15$$

...the product is positive!

But, on the other hand, when only **ONE** factor is negative...

$$(-3)(+5) = -15$$
$$(+3)(-5) = -15$$

...the product is negative and it **DOESN'T MATTER WHICH FACTOR IS NEGATIVE!**

—26—

You said the same rules apply to division?

YES! Again, if we have two negatives the quotient is **positive!**

$$-28/-7=+4$$

But like before, **ONE** negative and the quotient is negative and it doesn't matter **which** number is negative, the numerator **OR** the denominator!

$$-28/+7=-4$$
$$+28/-7=-4$$

I think my brain **just** blew a gasket!

I'm totally perplexed Professor!

I was lost right after "Good morning students."

Hmmmmmm...I sense some mass confusion out there. Well, like I always say..."Desperate situations call for desperate measures."

Please excuse me for one moment class.

TA DA

"I'M M.C. PROFESSOR AND I'M HERE TO EXPLAIN MUL-TEEPLYING! DEE-VIDING! ENRICHING YOUR BRAIN!"

"WHEN YOU DO EITHER ONE JUST REMEMBER THIS RULE '2 NEGATIVES MAKE A POSITIVE' AND YOU WON'T BE NO FOOL!"

"IF BOTH FACTORS ARE NEGATIVES WHEN WE TRY TO MUL-TEEPLY, THE PRODUCT IS A POSITIVE NO MATTER WHAT WE TRY!"

"IF ONLY ONE FACTOR'S NEGATIVE IT DON'T MATTER WHICH THEN THE PRODUCT'S ALSO NEGATIVE LET ME FINISH OFF MY PITCH!"

BOOMA

BOOM BOOM

BOOM

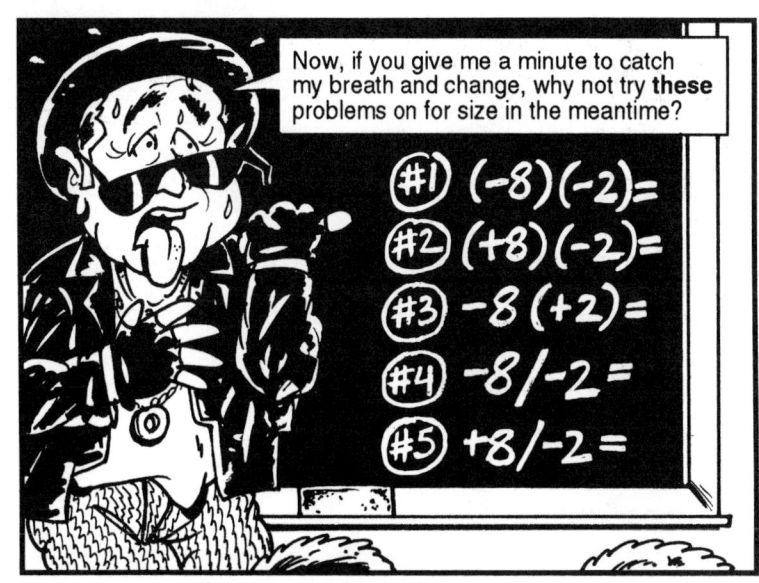

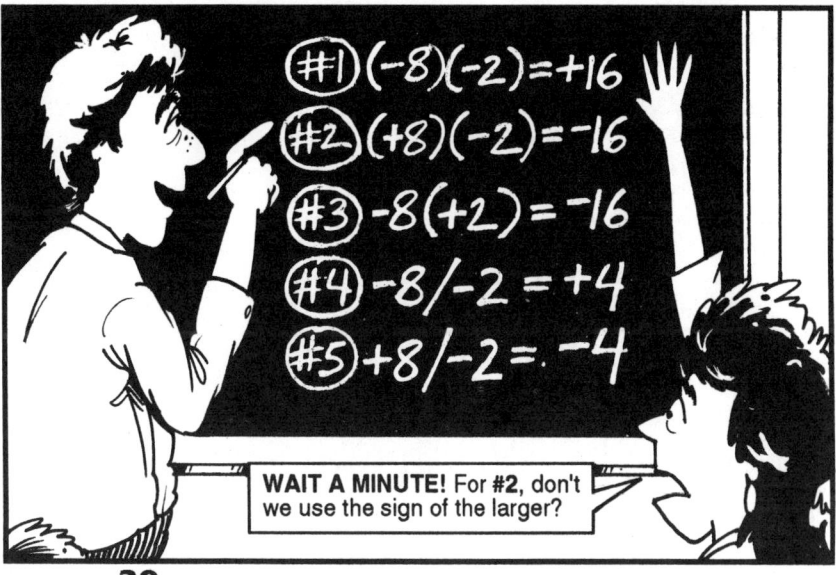

WORKING WITH DEE-VISION THE RULES ARE QUITE THE SAME THE QUOTIENT IS A POSITIVE IF TWO NEGS ARE IN THE GAME!

SCRATCHA SCRATCH SCRATCHA SCRATCH

AND LIKE BEFORE, ONE NEGATIVE WHICH ONE, DON'T MAKE A DIFF GIVES A QUOTIENT THAT'S A NEGATIVE. NOW CHECKOUT MY DEEJAY RIFF!

YEAH!! GET DOWN PROFESSOR!

Now, if you give me a minute to catch my breath and change, why not try **these** problems on for size in the meantime?

#1) $(-8)(-2)=$
#2) $(+8)(-2)=$
#3) $-8(+2)=$
#4) $-8/-2=$
#5) $+8/-2=$

TWO EXHAUSTED MINUTES LATER

OK! Now that I've returned to normal, relatively speaking, **anybody** wanna give these a try? How about you?

#1) $(-8)(-2)=+16$
#2) $(+8)(-2)=-16$
#3) $-8(+2)=-16$
#4) $-8/-2=+4$
#5) $+8/-2=-4$

WAIT A MINUTE! For **#2**, don't we use the sign of the larger?

Don't you remember? Only when you combine... we're multiplying now!
#1 (-8)(-2)=+16
#2 (+8)(-2)=-16
#3 -8(+2)=-16
#4 -8/-2 = +4
#5 +8/-2= -4
YEAH! Go back and check out page 8!
Here students, try some more. Professor Weissman says **these** are a bit more challenging!
#1 -5/+5=
#2 -2/+8=
#3 -3-4-5=
#4 0/-5=
#5 -5/0=
Now, let me go over each answer and explain how we arrived there.
#1 -5/+5=-1
The answer to #1 is -1.
I got 0 Professor!
If you **combine** -5 and +5 you get 0...we're dividing here.
#2 -2/+8=-1/4
For this one...we come up with -1/4.
Shouldn't the answer be **-4**?
You divided the **8** by the **2**! It's just the opposite!
#3 -3-4-5=-12
The correct answer is -12. Remember that this is **not** multiplication. This is a **combining** problem.
#4 0/-5=0
In a case like this, a **zero** divided by any **non-zero** number equals zero.
-0 or +0?
Just 0! Zero has no sign.
#5 -5/0=?
UNDEFINED
This is tricky. We **can't** divide by zero. The answer is said to be **"UNDEFINED."** Let's elaborate.

Elaborate away Professor!

I'll tell you what I will do. Let me tell you a little story...

Oh goody...make me some warm milk...then you'll tuck me in...

What is this...a wisecracker convention? Just listen to this story about my first job!

GRATEFUL DEAD

During high school, I used to eat...er...I mean deliver pizzas in my spare time. Every Saturday night...just like clockwork...I'd deliver a pizza to Elmo Needleman's house for the weekly poker game with his buddies. Even back then, in high school, the budding Math Professor in me would surface rather frequently.

One week I showed up with the pizza pie as usual and Elmo had three friends over.

Hey Mr. Needleman...check it out. **8** slices of pizza...**4** people. **8÷4=2**. That's **2** slices each!

Gee...thanks kid! We **never** would have figured that out without you!

The following week...Elmo had just one of the guys over!

Hmmmmm...only **2** people today. Let's see... **8** slices...**2** people. **8÷2=4**. **4** slices each!

You don't get **out** much...do ya kid?

Yeah! Keep **this** up and who knows... you **might** wind up being (har har) a Math Professor one day (har har)!

—31—

A week later...poor Elmo was there all alone playing solitaire for money...a most pathetic sight..
Geez Mr. Needleman...where **is** everybody?
Search me kid! **MAYBE YOU GUYS SHOULD MAKE BETTER PIZZA!**
Hey...we use the finest artificial ingredients money can buy...now as I was saying...
OK...we got **8** slices... **1** guy! **8÷1=8**. HEY! Are **you** gonna eat all **8** slices? Need any help?
Please kid... **NOT THE MATH AGAIN!**
Finally...the following week, I showed up with my pizza and NO ONE was home.
Hey...no one's there. **8** slices... **zero** people. **8÷0=?** Huh... What the? How can you divide **8** slices by **0** people?
HOLY SMOKES! Wait a minute! I just remembered what my math teacher told us. **YOU CAN'T DIVIDE BY ZERO! 8÷0** is said to be **UNDEFINED**. I'll have to bring the pizza back...undivided!
Well class...**that's** my little story. Are there any questions?
ZZZZZZ
SNXXXX
ZZZZ
ZZZ
ZZZZ
SNXXXX
(Sigh) I wish my class's attention was "**undivided**."
Let's break and try some more problems.

PROBLEM SET #2
See Answer Section in Back

214. The answer that you get when you multiply numbers is called the ....

215. The numbers that you multiply are called ....

FIND THE PRODUCTS.

216. (+8) (+7)

217. (-6)(-5)

218. (-8)(+10)

219. (0)(-7)

220. (+9)(-1)

221. -8(-3)

222. (-1)(-2)(-3)

223. (-1)(-2)(-3)(-4)

224. (-1)(-1)(-1)(-1)(-1)

225. -4(+3)(+2)(7)

226. (-1)(+2)(-3)(+4)

227. (-1)(+2)(-3)(+4)(-5)

228. The result to a division problem is called the .... and the number you are dividing by is called the ....

DIVIDE

229. +6 / +2

230. +2 / +6

231. -12 / -4

232. +12 / -4

233. -4 / -12

234. 0 / 4

235. 0 / -4

236. 7 / 0

237. -3 / 0

238. -3 / +3

239. $\dfrac{-12}{-15}$

240. $\dfrac{+40}{-1}$

SIMPLIFY EACH PROBLEM

241. $\dfrac{-2-4}{5-11}$

242. $\dfrac{(-2)(-4)}{-1-2-3}$

243. $\dfrac{9-(-3)}{-9(3)}$

244. Division by ..... is not allowed.

WHAT VALUE CAN X NOT BE ?

245. 12 / X

246. $\dfrac{12}{X-4}$

247. $\dfrac{12}{X+5}$

FIND THE MISSING NUMBER.

248. (-3)(?) = 12

249. (+15)(?) = -30

-33-

250. (?)(-8) = 0

251. (-9)(?) = +81

252. -10 / ? = +2

253. -10 / ? = +1

254. -10 / ? = -1

255. -10 / ? = 1/5

SIMPLIFY EACH EXPRESSION. SPLIT INTO TERMS FIRST.

256. -3 + 2(-6)

257. -3(+2) -6

258. -1 -2 -3 -4

259. -1(-2) -3(-4)

260. -1 -2(-3) -4

261. -1 -2 -3(-4)

262. (-1)(-2)(-3) -4

263. (-2.3)(-5.7)

264. (-0.2)(-0.2)(-0.2)

265. (-1/2)(-1/3)

266. -10(-0.7)

267. 100(-3.56)

268. $\dfrac{(-4) - (-3)}{(-2) - (+7)}$

269. $\dfrac{(+9) - (-1)}{(-1) - (+9)}$

270. $\dfrac{-1 \quad -1}{(-1)(-1)}$

A USEFUL RULE FOR THE SIGN OF A PRODUCT IS TO COUNT THE AMOUNT OF NEGATIVE FACTORS. IF YOU HAVE AN EVEN AMOUNT OF NEGATIVES THEN THE RESULT IS POSITIVE, OTHERWISE, FOR AN ODD AMOUNT, THE RESULT IS NEGATIVE.

IS THE ANSWER POSITIVE OR NEGATIVE ? JUST GIVE THE SIGN OF THE ANSWERS.

271. (15)(-8)(-9.8)(+4)

272. (+68)(-8)(-100)(-5)

273. (-10)(70)(89)(234)

A NEGATIVE ON THE LEFT OF PARENTHESES MEANS CHANGE THE SIGNS OF ALL TERMS INSIDE THE PARENTHESES.

$$- (7 - 5 + 3) = -7 +5 -3$$

This is the same as multiplying each term inside by -1. In fact, we can "squeeze" a 1 between the NEGATIVE and the LEFT PARENTHESIS and multiply each term inside by this negative one.

$$-1(7 - 5 + 3) = -7 +5 -3$$

The results are the same !!

When we multiply each term inside parentheses by the number outside we are DISTRIBUTING.

DISTRIBUTE

274. 2(x + y)

275. -3(a + b)

276. 7(2x -1)

277. -5(3x + 2)

278. 4(a + b - c)

279. -6(4x -1)

280. 2x(3y -1)

281. -3a(5 - 2b)

282. 5(2a -3b +4c)

283. -2(3x -5y + z)

284. 5(2x -1) -3(4x +1)

285. 5(3x +2) - (7 - 3x)

286. 2(2a - b) -4(a -b)

287. 2a(3b -5) -2b(3a +5)

288. 5(6-3x) -x(7y -1)

289. Subtract 4(3x-y) from 5(y-x)

290. Find the difference of 5x and 3(x-7)

291. Find the difference of 0 and 4(x-7)

292. What is 2(x-7y) increased by 7(y-5)?

293. The length of a rectangle is x and the width is x-5. What is the perimeter?

294. What is the value of x nickels and 2x-3 dimes?

AN IMPORTANT FORMULA IS RATE X TIME = DISTANCE.

295. A man in a car drives at 50 miles per hour. How far will he have travelled in 7 hours?

296. A train leaves a station travelling at x-5 miles per hour. How far will it have gone in 7 hours?

ANOTHER IMPORTANT FORMULA IS: RATE X PRINCIPAL=INTEREST

297. If you invest $600 at 8%, what is the interest?

298. If you invest $X at 4%, what is the interest?

299. If you invest $(300-X) at 4%, what is the interest?

300. If you invest $(2x-300) at 6%, what is the interest?

301. If you invest $X at 7% and $(3x -400) at 6%, what is the total interest?

302. How much money would 12 $5 tickets and 8 $7 tickets cost?

303. How much money would x $5 tickets and y $7 tickets cost?

304. How much would x $5 tickets and 2x-1 $7 tickets cost?

OK class...let's move on. We've covered the four basic operations. **Now** it's time to discuss powers and exponents.
ADDITION
SUBTRACTION
MULTIPLICATION
DIVISION
POWERS/EXPONENTS
Some terminology we'll use are as follows.
4^3
EXPONENT
BASE
Go no further Prof! The answer is **12**!
Well...you're thinking three **4's**... that's good...but you're **ADDING** them.
4^3
EXPONENT
BASE
No...I didn't...I **multiplied**! 3x4=12.
Multiplication **IS** repeated addition! **3x4=4+4+4=12** but exponents mean repeated **multiplication**. Let me show you!
Enlighten us oh great guru of Algebra.
By definition, the exponent, **3**, tells us to write the base, **4**, as a **FACTOR** three times.
$4^3 = 4 \cdot 4 \cdot 4$
EXPONENT
BASE
FACTORS? I remember. So we're multiplying the three **4's**. Let me give it a shot!
Shoot away!
$4^3 = 4 \cdot 4 \cdot 4$
$16 \cdot 4$
OK...let's see! **4x4=16**. Then...**16x4=** uh...uh... anybody got a calculator?
HA
HO
C'mon now...**16x4=64**!
Oh...oh yeah! OK, so therefore **4^3=64.**
$4^3 = 4 \cdot 4 \cdot 4$
$16 \cdot 4$
64

—37—

Right! Let's try another problem.
3^4
That's 64! We just did this one!
No! Look! The Professor switched the base and the exponent.
Correct! Here we have 4 factors of the base 3!
$3^4=3\cdot3\cdot3\cdot3$
Alright...so we have 3x3=9. Then 9x3=27 and finally 27x3=81! Right?
Yes! But it might be easier to say 9x9=81. See what I mean?
$3^4=3\cdot3\cdot3\cdot3$ → $9\cdot9$ → 81
Man...I finally got one right and I'm still wrong!
Relax...what's that expression? "Don't have a moose?" Now, before we move along, it's very important to be able to identify the base.
$(-3)^4$ BASE | -3^4 BASE
In this first example, the base is everything you see in the parentheses. So you actually have 4 factors of -3. We then multiply as usual.
$(-3)^4=(-3)(-3)(-3)(-3)$ → $(+9)\ (+9)$ → 81
Now over here, without parentheses, the negative is NOT part of the base. Ignore the negative sign for the time being. Do the multiplication as usual, then carry the negative down to your product.
$-3^4=-3\cdot3\cdot3\cdot3$ → $9\cdot9$ → -81

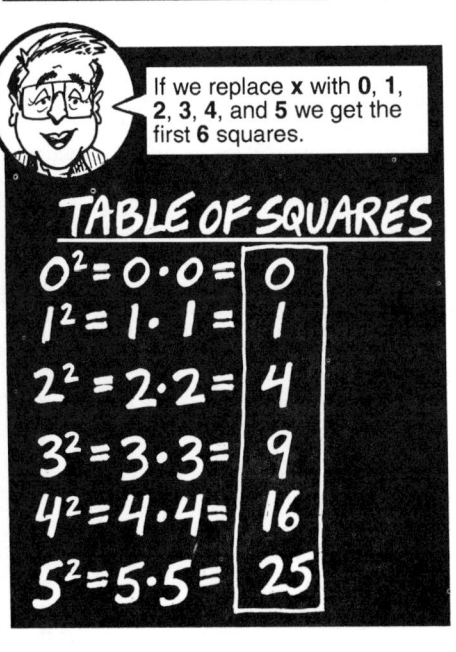

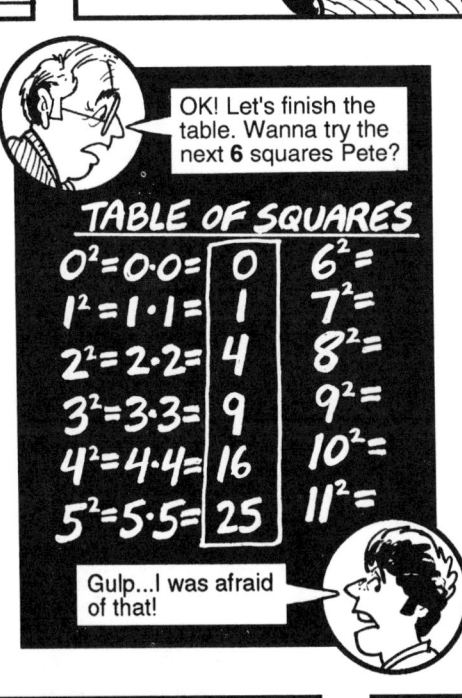

Panel 1: Let me generalize a bit. Here the exponent is **5**. We read this as "**x to the fifth power**."

$$x^5$$

Panel 2: In this example the base, **x**, is written as a factor **5** times.

$$x^5 = x \cdot x \cdot x \cdot x \cdot x$$

Panel 3:
$$x^2 = x \cdot x$$
"X SQUARED"

$$x^3 = x \cdot x \cdot x$$
"X CUBED"

Now in these examples when the exponent is **2** we use the word "**SQUARED**" and when the exponent is 3 we use the word "**CUBED**."

Panel 4: So what are the **answers** Prof?

We can't go any further **unless** we know the value of **x**. Let me show you.

Panel 5: If we replace **x** with **0, 1, 2, 3, 4,** and **5** we get the first 6 squares.

TABLE OF SQUARES
$0^2 = 0 \cdot 0 = 0$
$1^2 = 1 \cdot 1 = 1$
$2^2 = 2 \cdot 2 = 4$
$3^2 = 3 \cdot 3 = 9$
$4^2 = 4 \cdot 4 = 16$
$5^2 = 5 \cdot 5 = 25$

Panel 6: OK! Let's finish the table. Wanna try the next 6 squares Pete?

TABLE OF SQUARES
$0^2 = 0 \cdot 0 = 0$ $6^2 =$
$1^2 = 1 \cdot 1 = 1$ $7^2 =$
$2^2 = 2 \cdot 2 = 4$ $8^2 =$
$3^2 = 3 \cdot 3 = 9$ $9^2 =$
$4^2 = 4 \cdot 4 = 16$ $10^2 =$
$5^2 = 5 \cdot 5 = 25$ $11^2 =$

Gulp...I was afraid of that!

Panel 7: Ummmmm...let's see. Is **this** right Professor?

$6^2 = 6 \cdot 6 = 36$
$7^2 = 7 \cdot 7 = 49$
$8^2 = 8 \cdot 8 = 64$
$9^2 = 9 \cdot 9 = 81$
$10^2 = 10 \cdot 10 = 100$
$11^2 = 11 \cdot 11 = 121$

Panel 8: Correct! You see there was nothing to be afraid of. Now let's move to the **Table of Cubes**.

TABLE OF CUBES
$0^3 = 0 \cdot 0 \cdot 0 = 0$
$1^3 = 1 \cdot 1 \cdot 1 = 1$
$2^3 = 2 \cdot 2 \cdot 2 = 8$
$3^3 = 3 \cdot 3 \cdot 3 = 27$

That must be next to the **Table of Beverages**. (Nyuk)

Panel 9: OK wiseguy...finish the table!!

TABLE OF CUBES
$0^3 = 0 \cdot 0 \cdot 0 = 0$
$1^3 = 1 \cdot 1 \cdot 1 = 1$
$2^3 = 2 \cdot 2 \cdot 2 = 8$
$3^3 = 3 \cdot 3 \cdot 3 = 27$
$4^3 =$
$5^3 =$
$6^3 =$

Wiseguy? You better change that to **smart guy** Professor!

TABLE OF CUBES
$0^3 = 0 \cdot 0 \cdot 0 = 0$
$1^3 = 1 \cdot 1 \cdot 1 = 1$
$2^3 = 2 \cdot 2 \cdot 2 = 8$
$3^3 = 3 \cdot 3 \cdot 3 = 27$
$4^3 = 4 \cdot 4 \cdot 4 = 64$
$5^3 = 5 \cdot 5 \cdot 5 = 125$
$6^3 = 6 \cdot 6 \cdot 6 = 216$

OK...OK...beginner's luck! Now class... remember those **squares** and **cubes**. We'll be using them later on.

All's well and good Professor Weissman but are we **really** going to **use** exponents in our everyday lives?

Exponents are simply a convenient way of writing very large **OR** very small numbers. **You** might not be dealing with **those** kinds of numbers now but who knows? For instance ...

Cogsworth ... you **idiot**! The next time I ask you to calculate earth's distance to the Andromeda Galaxy ... a simple 10^{48} will **suffice**!

Geez Dr. Powers, I just **love** your new scale. It says I **only** weigh a mere 7^3 pounds. Incidentally, the 3 on your scale is kinda small and raised.

Seriously though, the use of exponents can be better explained if we **first** talk about the **Powers of 10**. Of course, we can use any number for the base. However, **10** is very special. Let's see why!

$10^4 = 10 \cdot 10 \cdot 10 \cdot 10$
$ 100 \cdot 100$
$ 10000$

With **10** as the base, the exponent tells us "how **many** zeros." So 10^4 means a 1 followed by 4 zeros.

In that case, what would 10^6 mean?

10^6

I got that one Prof!

$10^6 = 1000000$
A 1 followed by 6 zeros. Hey, I just made my first million!
Right! Study the Powers of Ten Table a moment. Do you notice a pattern. See how a 1 is always followed by the number of zeros that the exponent indicates.
$10^1 = 10$
$10^2 = 100$
$10^3 = 1000$
$10^4 = 10000$
$10^5 = 100000$
$10^6 = 1000000$
Patterns are important in life. Using our number pattern, what would you guess 10^0 to be?
$10^0 = ?$
$10^1 = 10$
$10^2 = 100$
$10^3 = 1000$
$10^4 = 10000$
$10^5 = 100000$
$10^6 = 1000000$
Zero! 10^0 is zero right?
$10^0 = 1$
$10^1 = 10$
$10^2 = 100$
$10^3 = 1000$
No wait! It would be a 1 followed by zero zeros. That's a 1!
Huh! 10^0 is 1? That doesn't look right?
Oh...but it IS right! He followed the pattern. Now, let me go one step further. ANY base (except 0) to the zero power is 1! See...
$7^0 = 1$
$12^0 = 1$
$b^0 = 1$
I don't get it! EVERYTHING with 0 as an exponent is 1?
YES! Except zero! 0^0 is said to be "UNDEFINED."
0^0 = UNDEFINED
There's that old "UNDEFINED" again.
Professor Weissman's
BELIEVE IT OR NOT
Any Base (Except 0) to the 0 Power is 1!
We'll explain this amazing fact later on.
Let's start from the beginning. I'll be back in a sec class.

Hereyee...Hereyeee! All rise! The Honorable Judge Weissman presiding!

Here come da judge! Here come da judge! Boy... is the Professor committed to his job!

He should be **"committed"** alright!

Now.. let's see here. This **isn't** the book I was looking for but check this out anyway!

THE AFTERMATH OF THE ROMAN EMPIRE

You think learning math is tough **now**? Look what students in Ancient Rome **had** to contend with!

MULTIPLY MCXVIII · LXIV

Yikes! No wonder the Roman Empire fell!

Oh goody... **here's** the book I was looking for.

LAWS ~OF~ EXPONENTS

OK! LAW #1 (aka: The Product Rule)— If it is deemed necessary for the party of the first part (you) to multiply numbers with the same base...then the party of the first part will under the penalty of law, retain the same base without alteration (keep the same base—don't change it in any way). The party of the first part (you again) must use that same base for your product. The value of the party of the second part (the exponent) for your product will be the sum of the given exponents. The following law is only applicable (you can only use it) when you are:
1. Multiplying numbers and
2. The numbers have the same base

PARTY? Did someone say **party**?

Could you say that in the King's English your honor?

—42—

"Why soitenly! This is **LAW #1** more clearly stated."

LAW #1 OF EXPONENTS
"The Product Rule"

To Multiply Numbers with the Same Base, Keep the Base & Add the Exponents

"Here class! Check out this example of **LAW #1** in action."

LAW #1 – "The Product Rule"
$$x^a \cdot x^b = x^{a+b}$$
$$x^3 \cdot x^5 = x^{3+5}$$
$$= x^8$$

The dot indicates we're multiplying and it's obvious we have the same base, **x**. Thus, all we do is add the exponents and use the same base giving us x^8.

"What's the product in **this** example?"

$$x^4 \cdot x^6 =$$

"That's x^{10}! I understand 4+6=10 but why do we **add** the exponents?"

"Good question! Let's see **why LAW #1** applies."

"The definition of exponents will explain **LAW #1**. Let's write x^4 and x^6 using factors."

$$x^4 \cdot x^6 =$$
$$\underbrace{xxxx}\;\underbrace{xxxxxx}$$

"See how we are actually **combining** or **adding** factors. Now, what have we got?"

"A perfect bowling score?"

HA HO HA HE HO HE

"True, but we **also** end up getting **10** factors of **x**. Now can you see why we added the **4** and the **6**? Here are some more problems. Try to apply **LAW #1**."

#1) $x^6 \cdot x^2 =$
#2) $x^3 \cdot x =$
#3) $2^5 \cdot 2^3 =$
#4) $3^4 \cdot 3 =$
#5) $(3a)^2 (3a)^3 =$
#6) $3a^5 \cdot 2b^4 =$
#7) $x^3 \cdot y^2 =$
#8) $x^3 + x^2 =$
#9) $2^3 \cdot 3^2 =$

"And to our readers out there, try solving these as well before turning the page."

–43–

Panel 1 (Professor at blackboard): Let's put in the answers. Any questions?

#1) $x^6 \cdot x^2 = x^8$
#2) $x^3 \cdot x = x^4$
#3) $2^5 \cdot 2^3 = 2^8$
#4) $3^4 \cdot 3 = 3^5$
#5) $(3a)^2 (3a)^3 = (3a)^5$
#6) $3a^5 \cdot 2b^4 = 6a^5 b^4$
#7) $x^3 \cdot y^2 = x^3 \cdot y^2$
#8) $x^3 + x^2 = x^3 + x^2$
#9) $2^3 \cdot 3^2 = 2^3 \cdot 3^2$

Panel 2: For #1, I had x^{12} but I see my mistake. I should have **added** the exponents.
#1) $x^6 \cdot x^2 = x^8$

Panel 3: For #2, how'd you get x^4?
#2) $x^3 \cdot x = x^4$

Panel 4: When **no** exponent is written, an exponent of **1** is assumed.
#2) $x^3 \cdot x^{1} = x^4$

Panel 5: Remember LAW #1 says to keep the base. The base of the product is **ALSO** 2.

Panel 6: For #4, here again we have the assumed exponent of **1**.
#4) $3^4 \cdot 3^{1} = 3^5$
Did you keep the base?

Panel 7: #3) $2^5 \cdot 2^3 = 2^8$
For question #3, isn't the answer 4^8?

Panel 8: Hey Prof! For #5, I have $3a^5$ without the parentheses. Is **that** OK?

Panel 9: Nope! In your answer the **3** **isn't** part of the base like it **is** in the two numbers we're multiplying.

$(3a)^5$
BASE

With parentheses, the base is **3a** and the exponent is **5**.

Panel 10: **Without** parentheses, the exponent is still **5** but the base is just **a**.

$3a^5$
BASE

Then what's the 3 in that answer?

Panel 11: In this case, the 3 would be considered a **coefficient**. Don't you remember we discussed that at the bottom of page 14?

COEFFICIENT
$3a^5$ ← BASE

—44—

Now, in this example, even though the bases, **a** and **b**, are different, we can still simplify the problem a bit by multiplying the coefficients, the **3** and **2**, giving us **6**.

#6 $3a^5 \cdot 2b^4 = 6a^5b^4$

Was I lazy or did I have another reason? You tell me!

Right! On #7 the bases are different. Remember **LAW #1** only refers to multiplying numbers with the same base.

#7 $x^3 \cdot y^2 = x^3 \cdot y^2$
↑   ↑
DIFFERENT BASES

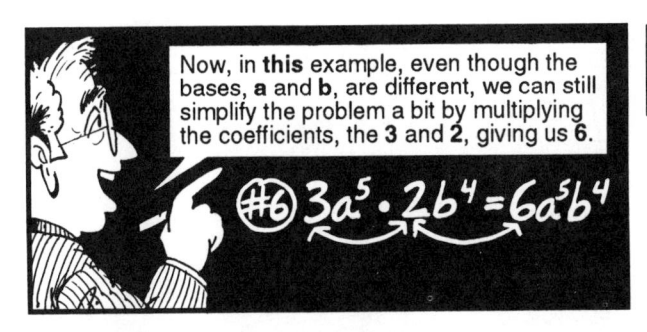

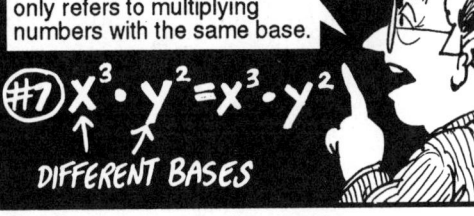

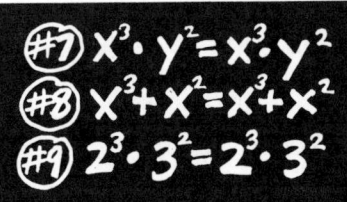

#7 $x^3 \cdot y^2 = x^3 \cdot y^2$
#8 $x^3 + x^2 = x^3 + x^2$
#9 $2^3 \cdot 3^2 = 2^3 \cdot 3^2$

Oh yeah...I see. But Professor, you **didn't** do #7, #8, and #9.

I got it! They don't fit the conditions for **LAW #1**!

The Product Rule doesn't apply to **this** example either. We're **not** multiplying here...we're **adding**.

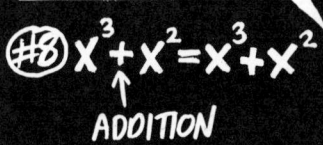

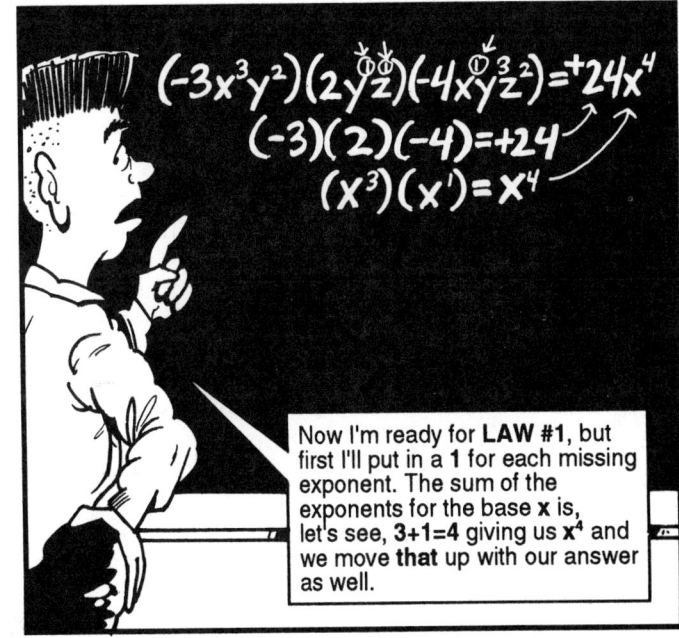

#8 $x^3 + x^2 = x^3 + x^2$
↑
ADDITION

Now once again, in our last example, **LAW #1** doesn't apply. However, we **can** simplify it using the definition of exponents. Look!

#9 $2^3 \cdot 3^2$
$2 \cdot 2 \cdot 2 \cdot 3 \cdot 3$
$8 \cdot 9$
72

Now class, let me **really** blow your minds with this problem. Anybody feel up to **this** one?

$(-3x^3y^2)(2yz)(-4xy^3z^2) =$

OY!

You **must** be dyslexic! The word is **YO!** I'll tackle it Prof!

$(-3x^3y^2)(2yz)(-4xy^3z^2) =$

$(-3x^3y^2)(2yz)(-4xy^3z^2) = {}^+24$
$(-3)(2)(-4) = +24$

First I'll multiply the three coefficients, the **-3**, the **2**, and the **-4** so at **least** we know **+24** is part of the answer.

$(-3x^3y^2)(2yz)(-4xy^3z^2) = {}^+24x^4$
$(-3)(2)(-4) = +24$
$(x^3)(x^1) = x^4$

Now I'm ready for **LAW #1**, but first I'll put in a **1** for each missing exponent. The sum of the exponents for the base **x** is, let's see, 3+1=4 giving us x^4 and we move **that** up with our answer as well.

—45—

Panel 1 (blackboard):

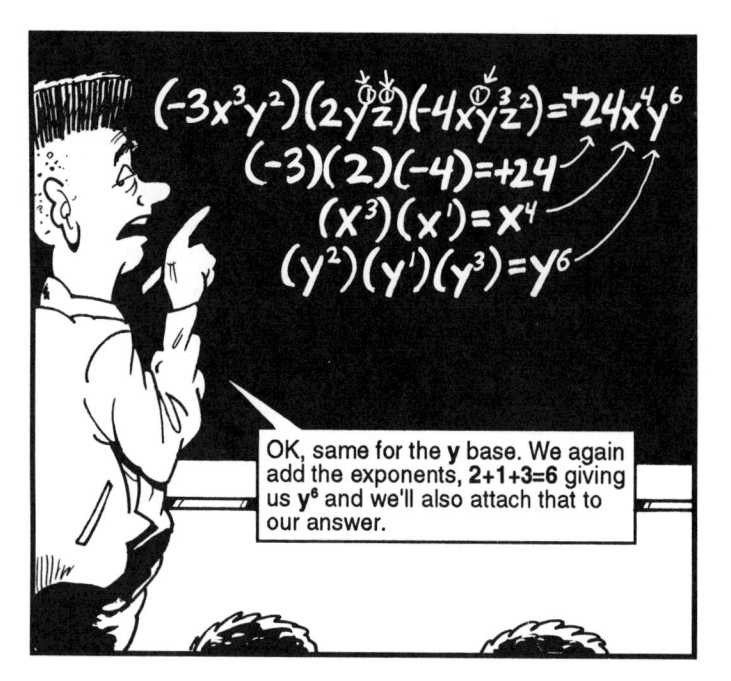

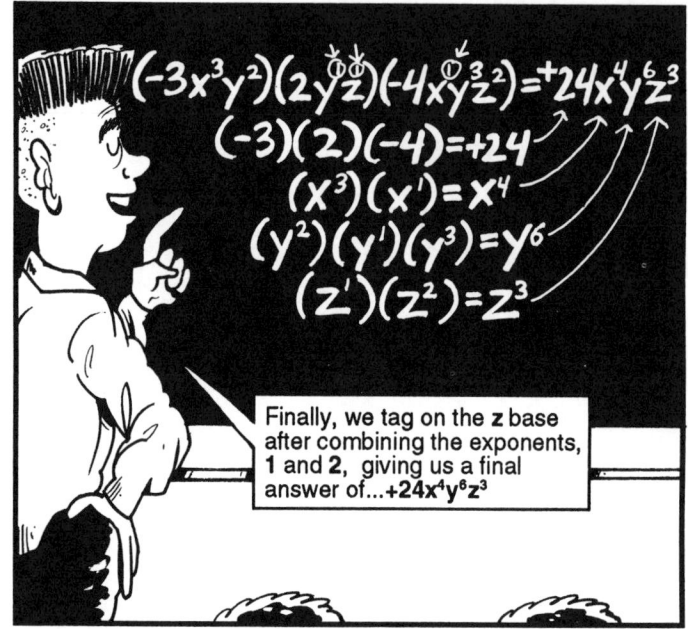

$$(-3x^3y^2)(2yz)(-4xy^3z^2) = +24x^4y^6$$
$$(-3)(2)(-4) = +24$$
$$(x^3)(x^1) = x^4$$
$$(y^2)(y^1)(y^3) = y^6$$

"OK, same for the **y** base. We again add the exponents, 2+1+3=6 giving us **y⁶** and we'll also attach that to our answer."

Panel 2 (blackboard):

$$(-3x^3y^2)(2yz)(-4xy^3z^2) = +24x^4y^6z^3$$
$$(-3)(2)(-4) = +24$$
$$(x^3)(x^1) = x^4$$
$$(y^2)(y^1)(y^3) = y^6$$
$$(z^1)(z^2) = z^3$$

"Finally, we tag on the **z** base after combining the exponents, **1** and **2**, giving us a final answer of... $+24x^4y^6z^3$"

Panel 3:

"BRAVO! Well done! I think we have a ringer here. Let's move on to **LAW #2**."

"How many of these laws **are** there?"

Panel 4:

"Don't worry...only **3** and **LAW #2**, The **Quotient Rule**, follows nicely from **LAW #1**. See what I mean?"

LAW #~~1~~ 2 OF EXPONENTS
"The ~~Product Rule~~"
Quôtient

To ~~Multiply~~ Divide Numbers with the Same Base,

Keep the Base &

~~Add~~ Subtract the Exponents

Panel 5:

"Let's try an example."

LAW #2 - "The Quotient Rule"

$$\frac{x^a}{x^b} = x^{a-b}$$

$$\frac{m^5}{m^3} = m^{5-3} = m^2$$

"Remember, we **keep** the common base and **subtract** the exponents. As you recall, for **LAW #1** or **The Product Rule**, I explained why the exponents are added. Can anybody volunteer why we **subtract** here?"

Panel 6:

"I guess we go back to the **definition of exponents** again...huh?"

$$\frac{m^5}{m^3} = \frac{mmmmm}{mmm}$$

"You guessed right. We have **5** factors of **m** in the numerator and **3** factors of **m** in the denominator. What now?"

—46—

CANCEL?

Righto! We can cancel **3** of the **m**'s in the numerator with the **3 m**'s from the denominator.

$$\frac{m^5}{m^3} = \frac{\cancel{m}\cancel{m}\cancel{m}mm}{\cancel{m}\cancel{m}\cancel{m}}$$

So when we're **dividing** we're **cancelling** or **subtracting** factors.

$$\frac{m^5}{m^3} = \frac{\cancel{m}\cancel{m}\cancel{m}mm}{\cancel{m}\cancel{m}\cancel{m}} = \frac{m^2}{1} = m^2$$

Simplifying we get **m²**. That's the same result we got when we applied **LAW #2** or **The Quotient Rule**.

Let's try some more problems folks.

#1) $\dfrac{y^8}{y^2} =$ **#2)** $\dfrac{f^5}{f} =$ **#3)** $\dfrac{8^5}{8^2} =$ **#4)** $\dfrac{-8c^5}{+2c^3} =$ **#5)** $\dfrac{b^3}{b^3} =$

#1) $\dfrac{y^8}{y^2} = y^6$ **#2)** $\dfrac{f^5}{f} = f^4$ **#3)** $\dfrac{8^5}{8^2} = 8^3$ **#4)** $\dfrac{-8c^5}{+2c^3} = 4c^2$ **#5)** $\dfrac{b^3}{b^3} = 0 \underline{\underline{OR}} 1$

Let's fill in the answers and elucidate each one.

You're a veritable cornucopia of information Prof!

Information is my middle name...well actually it's Irwin but I **prefer** Information. Now in our first example, we **keep** the base and **subtract** the exponent.

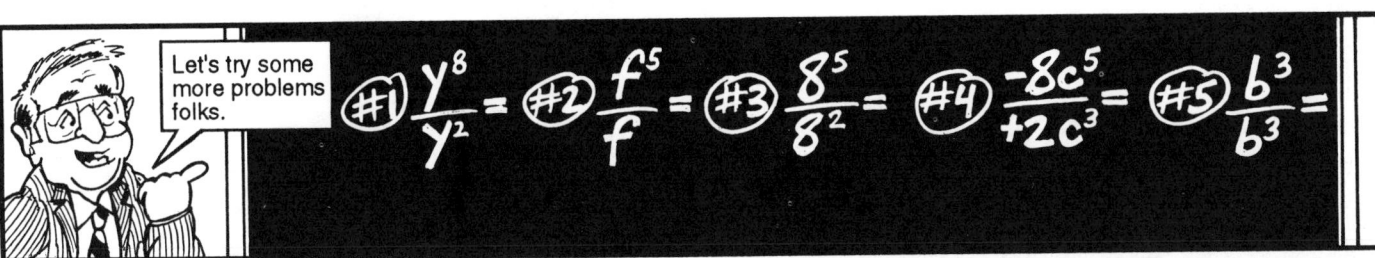

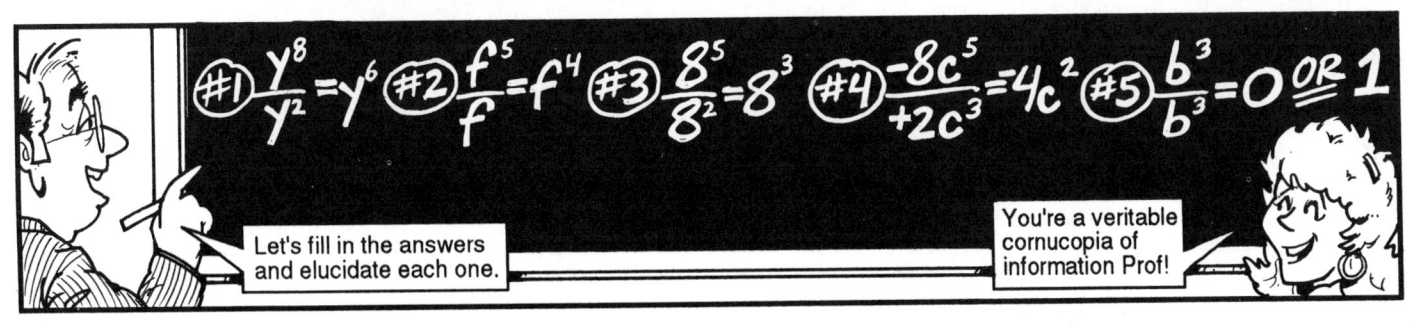

#1) $\dfrac{y^8}{y^2} = y^{8-2} = y^6$

For **#2**, fill in the assumed exponent.

#2) $\dfrac{f^5}{f^{①}} = f^{5-1} = f^4$

#3, business as usual.

#3) $\dfrac{8^5}{8^2} = 8^{5-2} = 8^3$

—47—

Huh! I got 1³. 8÷8=1... isn't it?
Did you forget already? The Prof said not to mess with that base. We have to keep the common base. Hang on to that 8!
#4 $\frac{-8c^5}{+2c^3} = {}^-4c^{5-3} = {}^-4c^2$
Well said! Now, in #4, we have to remember the -8 and +2 are NOT part of the base. Divide them as usual giving us a quotient of -4. Keep the common base, c, and subtract those exponents.
For our last example here, we can have two different answers.
#5 $\frac{b^3}{b^3} = b^0$ OR 1
I got b⁰. I understand that! I kept the common base and subtracted the exponents. Isn't b⁰ the only answer? What do you mean OR 1?
That's my "Believe it or Not." We can explain why b⁰=1 if we tackle the problem using the definition of exponents.
$\frac{b^3}{b^3} = \frac{bbb}{bbb}$
$\frac{b^3}{b^3} = \frac{bbb}{bbb} = \frac{1}{1} = 1$
Now cancel... then simplify. That's how we get 1!
USING LAW #2 "The Quotient Rule"
$\frac{b^3}{b^3} = b^0$
USING DEFINITION & CANCELLING
$\frac{b^3}{b^3} = 1$
Both problems are the same...
...but the answers differ depending on which approach we take.
So both answers must be eqivalent?
Yes...you're right! b⁰=1 and since b could be any number (except 0) that explains my "Believe it or Not."
$b^0 = 1$

$$\frac{14a^3b^5c^7}{-2a^3bc^2}=$$

Now, let's put that gray matter to work again. Take a few moments to work out **this** problem. Feel free to discuss it amongst each other.

Oh boy... I think it's **-7**!

Subtract those exponents, right?

Is it a^0 or just a?

b has no exponent... what gives?

Meanwhile...as the class discusses the division problem

Hmmmm... I could have done a **better** job explaining why $b^0=1$. I hope the class really understood that **b** could be **ANY** number!

TAP TAP

While Professor Weissman ponders deeply, he recalls, with great amusement, the fun time he and his colleagues enjoyed relaxing in the Math Teachers' Lounge the other day. The laughs were coming fast and furious.

TEACHERS LOUNGE

Let's go guys...**somebody give me a number!**

Alright! I go first!

12!

How about 4?

Oh yeah...7!

That's rich! What about 14?

Stop! Stop! I can't breathe!

Are you folks **crazy** or what? I always see you shouting out numbers at random and laughing hysterically... **what gives**?

-49-

Relax Rosie! There **is** a rational explanation.
The gang and I were simply fed up with telling those same old hilarious but long-winded jokes so we devised a new system. We assigned each of our jokes a number, so instead of telling the whole joke, we just shout out the corresponding number. It saves a lot of time.
C'mon...give it a try! Call out a number Rosie!
OK...sure let's see... how about... uh...uh... **6**!
I don't get it! You all shout out numbers and go into hysterics. I call out a number and get diddley! Wasn't **my** joke funny?
It's not that the **joke** wasn't funny Rosie! It's just the **WAY** you told it!
Suddenly... a dropped pencil startles Professor Weissman from his daydream and back to reality. Quick as a wink, he solves the division problem up on the board.
PLINK
$\frac{14a^3b^5c^7}{-2a^3b^1c^2} = -7a^0b^4c^5$
$\frac{14}{-2} = -7$
$a^{3-3} = a^0$
$b^{5-1} = b^4$
$c^{7-2} = c^5$
OK class...first we divide the coefficients giving us **-7**. Now, using **LAW #2**, I'll put in a **1** for the missing exponent. Then subtract the exponents for the three different bases, **a**, **b**, and **c** to get our answer.
Careful with those **a's**. $a^{3-3}=a^0$...**NOT a**! And what is a^0 equivalent to...to **1**! That means we can simplify our answer a bit more. Watch!
$\frac{14a^3b^5c^7}{-2a^3b^1c^2} = -7a^0b^4c^5$
MULTIPLY
EQUIVALENT
1
$-7b^4c^5$
If we replace a^0 with the **1** we can multiply the **-7** and the **1** and **still** end up with just the **-7**. So the final answer reads as follows.

Ok folks! Do you recall my "**Believe it or Not**" about any number (except **0**) to the **0** power equalling **1**...
Yeah so?
Let's go guys...somebody give me a number!
TEACHERS LOUNGE
Alright! I go first!
OK...so somebody give me a number!
4
9
13
Sold! Let's use the **13**!
...and I said we could use **any** number except **0** for the **b**?
$b^0=1$
In the original example, we learned **two** answers were possible. b^0 using "**The Quotient Rule**" and **1** using "**definition and cancelling**" and that **those** answers were equivalent.
$1=\frac{1}{1}=\frac{b\cdot b\cdot b}{b\cdot b\cdot b}=\frac{b^3}{b^3}=b^{3-3}=b^0$
EQUIVALENT
Now, if we replace all the **b**'s with **13**...look what happens. The 13^0 is also equivalent to **1**. This would have worked with **any** number we had selected so it demonstrates that **ANY NUMBER TO THE 0 POWER IS 1**!
$1=\frac{1}{1}=\frac{13\cdot 13\cdot 13}{13\cdot 13\cdot 13}=\frac{13^3}{13^3}=13^{3-3}=13^0$
EQUIVALENT
And the **only** exception to the rule is 0^0 because if we replaced the base with a **0**, we'd have a **0** in the denominator.
Remember, you can't divide by **0**! It's "**UNDEFINED.**"
$\frac{0^3}{0^3}$=UNDEFINED
Visualize how we could use "**The Quotient Rule**" in an actual situation. Pretend I'm an astronaut on my way to the distant star of **Algebra Centauri**.

Space Cadet Weissman! Check the navigational controls. What **distance** do we still have to travel before reaching our destination?
Checking Captain! **Hey**, we still have 10^{10} kilometers to go!
What's our **maximum** speed cadet?
10^4 kilometers per hour sir!
$-3x^4(5^2)ab^2$
ALGEBRA CENTAURI
So what's our **estimated time of arrival**?
I'm on the case sir!
MOTION PROBLEMS
Hey! This should help!
TIME= DISTANCE / RATE
I'll use our distance and rate to calculate the time. Hmmmmm... let's see, **LAW #2 of Exponents** says to **keep** the base and **subtract** those exponents!
$TIME = \frac{DISTANCE}{RATE}$
$TIME = \frac{10^{10}}{10^4}$
$TIME = 10^{10-4} = 10^6$
Well Cadet...
10^6 hours or a **1** followed by **6** zeros. That's a **million hours**, huh Captain?
IN DAYS PLEASE!
I'll divide a **24** hour day into our million hour total giving us...geez ...**41,667** days!
41666.666
Now in **years** Cadet!
OK, OK, now I'll divide a **365** day year into the **41,667** day total and wind up with, **holey moley**! **114 years**!
114.15616
Better make sure we have enough coffee!
Never mind **that**! I'm gonna check out the **toilet paper supply**!

Panel 1: Well, I hope you appreciated my little scenario using the **Quotient Rule** and that we've all learned something from it!

You mean besides the fact that **NASA** must have **really** relaxed their restrictions for becoming an astronaut?

HAR HA

Panel 2: Hey Prof...I **noticed** in your spaceship story...that you **rounded** the answers. Could you refresh my little Alge**brain** a bit about **rounding numbers**?

Panel 3: No sweat! We do a **lot** of rounding everyday, almost automatically. For example...take my tie.

No thanks!

Panel 4: Not **literally**... figuretively!

Either way, I **don't** want that tie!

Panel 5: NO! NO! Consider what I **paid** for it!

$3

Class, does this mean I paid **exactly** $3 for the tie?

Panel 6: No, **approximately** $3!

I suspect a lot less.

Well...what **could** I have paid for it when I say I paid $3?

Panel 7: $2.98?

$3.25?

Panel 8: Good answers! **Both** of those prices would become $3 when **rounded** to the nearest one dollar. Let me explain.

$2.98 | $3.25
↑   ↑
ROUNDING DIGIT

First, we have to decide to what **place** we will be rounding. For example, here we will round to the nearest **one** (dollar). The number we find there is called the **rounding digit**!

—53—

Panel 1:

Next, the number on its right is called the **decision digit**. Finally...

DECISION DIGIT
↓
$2.9**8** | $3.**2**5
↑
ROUNDING DIGIT

I remember now!

Panel 2:

If the decision digit is **5 or more**, we add **1** to the rounding digit, if it's **less than 5**, we leave the rounding digit alone.

$2.98 | $3.25
+1

Panel 3:

Right! Either way, we drop all the digits to the **right** of the rounding digit.

$2.98 $3.25
+1
___
$3

ROUNDS UP   ROUNDS DOWN
TO $3   TO $3
(from $2.98) (from $3.25)

Rounding **digits**! Decision **digits**! Drop the **digits**! Hey! **I dig it**!

All this talk of **digits**! I need my **Digitalis**!

Panel 4:

OK gang, settle down! Let's get **back** to those **exponents**! If you're all so positive that you've got those exponents down pat...let's see just how **positive** you are when the exponent is a **negative**. Anybody wanna take a crack at **this** one?

5^{-2}

Panel 5:

I know it's not -10!

Is it +25?

No wait... -25?

This is **exactly** what's been gnawing at me Professor! **All** those other problems you gave us using the Quotient Rule worked out nice and tidy. The exponent in the numerator was **always** larger or equal to the exponent in the denominator so when we subtracted them, the **new** exponent was always positive or zero. I suppose your answer, 5^{-2}, was most likely the result of dividing...let's say...5^2 by 5^4?

Panel 6:

You're right! The problem you mentioned **does** give my result but I'm **still** waiting to find out what 5^{-2} is?

5^{-2}

Geez Professor! I was **just** beginning to understand these here exponents and **now** you've gone and turned my world **upside down**!

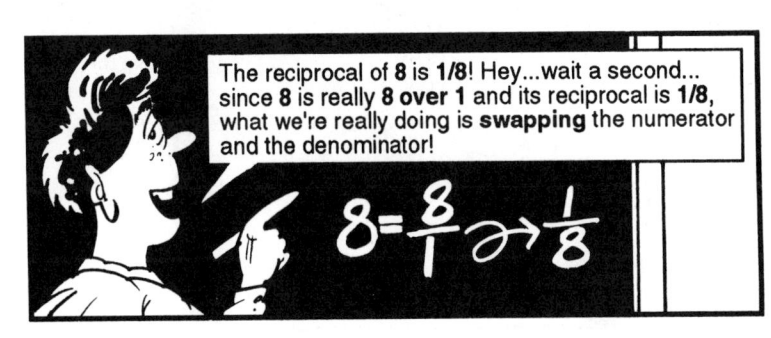

Panel 1: Ahhhhh! Did you say "**upside down**?" You're on the right track!

I'm on the **right** track? I don't even know where I'm going!

HA HA HAR

Panel 2: Please...allow me! In mathematics, we have a word for "**upside down**" and that word is **reciprocal**. Myron...have you got your dictionary handy?

RECIPROCAL

Panel 3: Right here Professor! Um...let's see now!
re·cip·ro·cate (ri sip′rə kāt), interchange. to cause to move alternately backward and forward.
Oh wait...here's reciprocal...
re·cip·ro·cal (ri sip′rə kəl), *Math.* the ratio of unity to a given quantity that by which the given quantity or expression is multiplied to produce unity: *The reciprocal of x is 1/x.*

DICTIONARY

Panel 4: Thanks Myron. The dictionary's example says "**the reciprocal of x is 1/x.**" What would the reciprocal of **8** be?

I got it!

Panel 5: The reciprocal of **8** is **1/8**! Hey...wait a second... since **8** is really **8 over 1** and its reciprocal is **1/8**, what we're really doing is **swapping** the numerator and the denominator!

$8 = \frac{8}{1} \rightarrow \frac{1}{8}$

Panel 6: **YES**! Now getting back to our original negative exponent problem, 5^{-2} is the **reciprocal** of 5^2, that is $1/5^2$.

$5^{-2} = \frac{1}{5^2} \text{ OR } \frac{1}{25}$

Panel 7: ...if you have a base with a **negative** exponent, flip the base to the denominator and **change** the sign of the exponent to a **positive**.

Panel 8: OK...OK! We've established that 5^{-2} equals $1/5^2$, but why?

Panel 9: So **that's** what you meant when you said I was on the right track! To find 5^{-2}, you turn 5^2 upside down?

Right! It might be easy to remember it this way...

That's what I used to tell my children Professor. If something bad or **negative** happens, turn it around and change it into something **positive**! Little did I know I was talking about **exponents**!

-55-

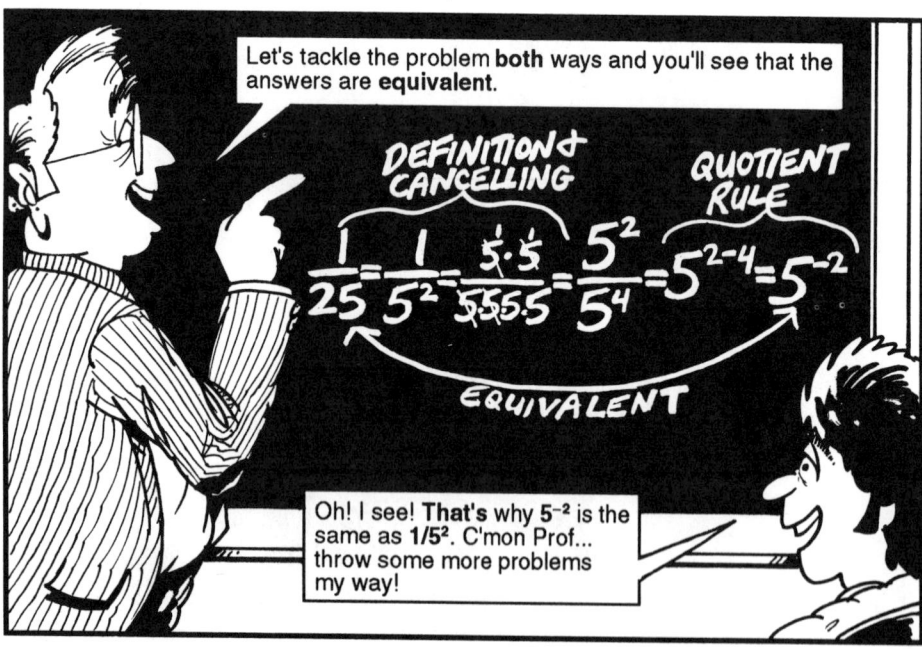

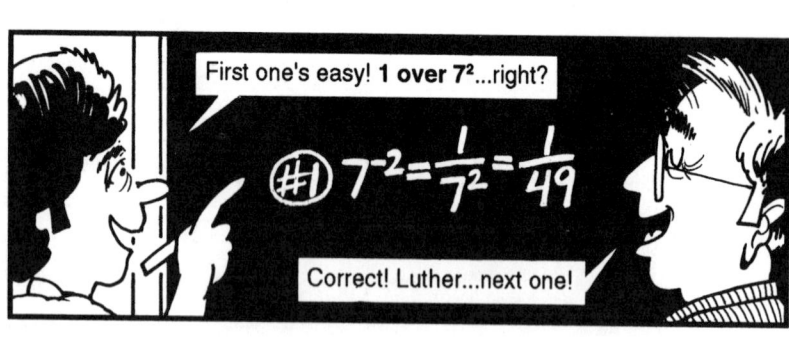

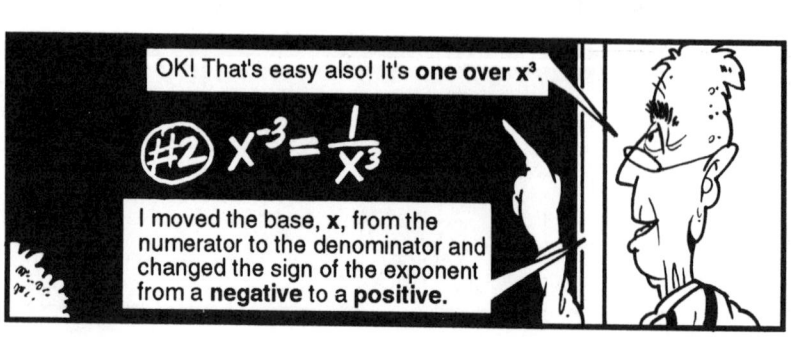

Panel 1: Let's explain it with the problem you mentioned before...and we'll use...

Panel 2: THE DEFINITION OF EXPONENTS!!

Panel 3: Sheesh...I must be getting predictable!

Panel 4: Let's tackle the problem **both** ways and you'll see that the answers are **equivalent**.

$$\underbrace{\frac{1}{25} = \frac{1}{5^2} = \frac{5 \cdot 5}{5 \cdot 5 \cdot 5 \cdot 5}}_{\text{DEFINITION \& CANCELLING}} = \underbrace{\frac{5^2}{5^4} = 5^{2-4} = 5^{-2}}_{\text{QUOTIENT RULE}}$$

EQUIVALENT

Oh! I see! **That's** why 5^{-2} is the same as $1/5^2$. C'mon Prof... throw some more problems my way!

Panel 5: Happy to oblige! Try the first one Mike!

#1) 7^{-2}
#2) x^{-3}
#3) $\frac{1}{y^{-5}}$

Panel 6: First one's easy! **1 over 7^2**...right?

#1) $7^{-2} = \frac{1}{7^2} = \frac{1}{49}$

Correct! Luther...next one!

Panel 7: Professor, can't we **just** say in a nutshell that when we move a base from the numerator to the denominator, the sign of the exponent changes?

Absolutely...and in fact, even when we move a base the **other** way, from the denominator to the numerator, the sign of the exponent **also** changes. Check out our **third** example.

#3) $\frac{1}{y^{-5}}$

Allow me Professor!

Panel 8: OK! That's easy also! It's **one over x^3**.

#2) $x^{-3} = \frac{1}{x^3}$

I moved the base, **x**, from the numerator to the denominator and changed the sign of the exponent from a **negative** to a **positive**.

—56—

–57–

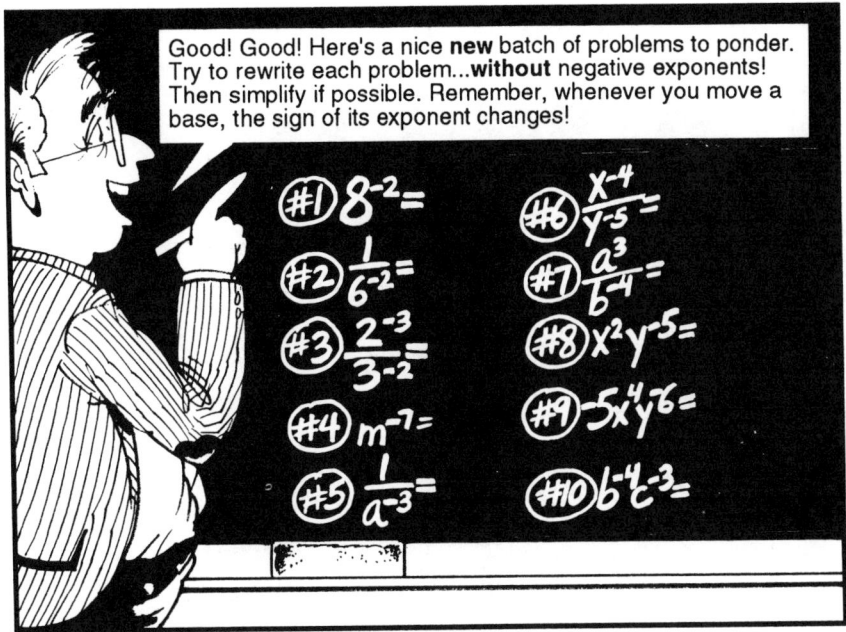

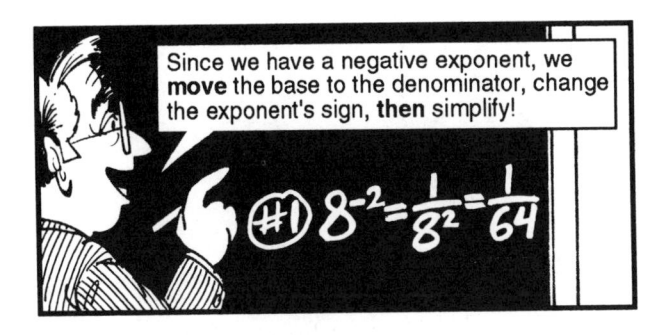

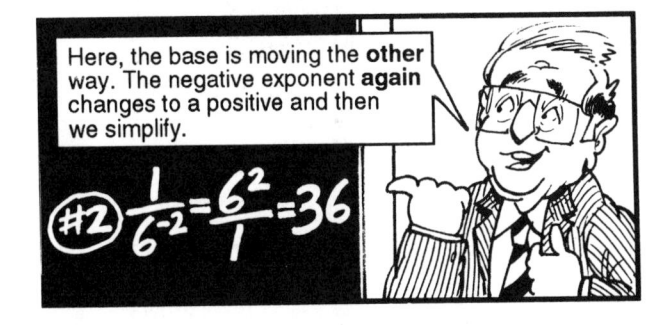

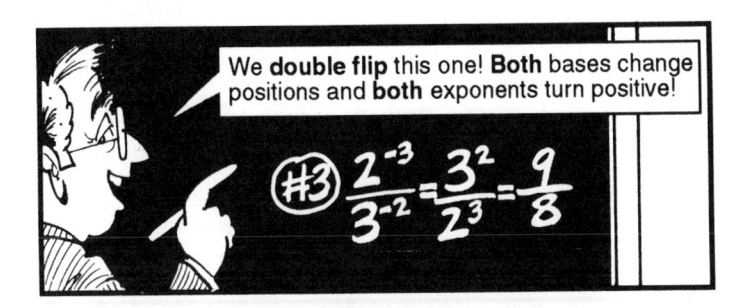

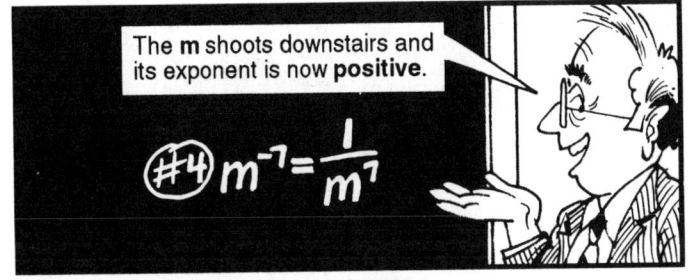

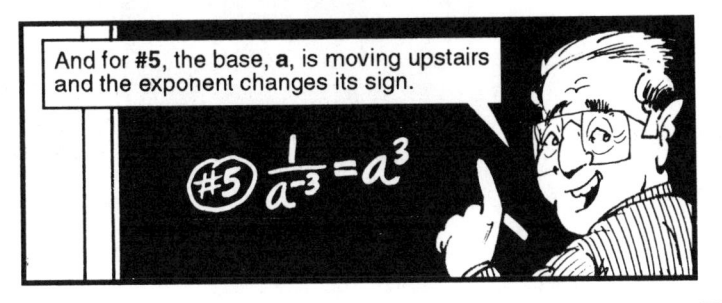

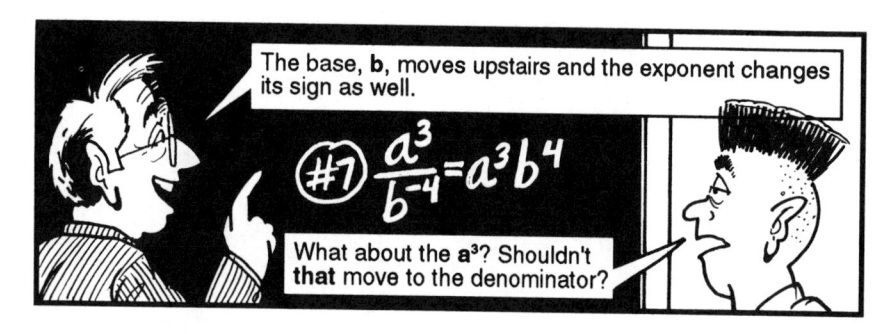

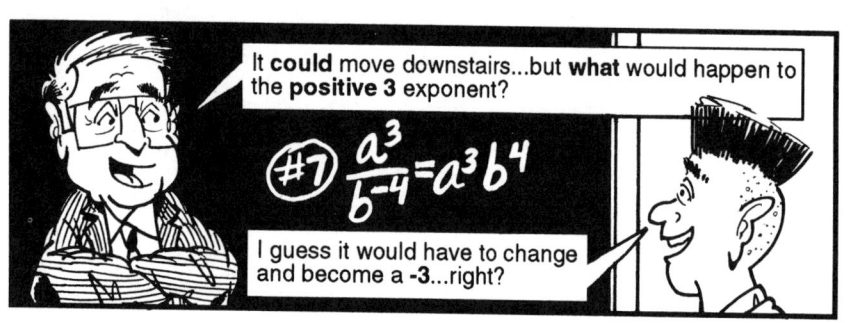

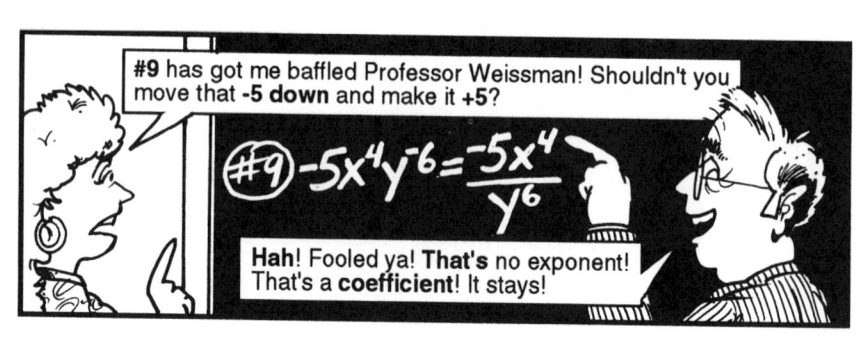

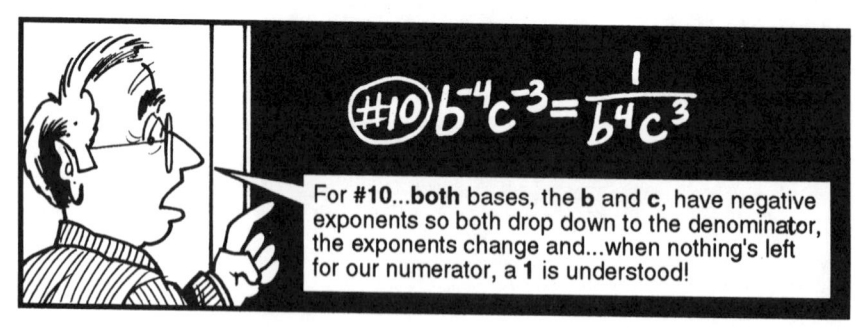

Panel 1: The base, **b**, moves upstairs and the exponent changes its sign as well.

#7 $\dfrac{a^3}{b^{-4}} = a^3 b^4$

What about the a^3? Shouldn't **that** move to the denominator?

Panel 2: Righto! But remember, I wanted the problems rewritten **without** negative exponents!

Panel 3: It **could** move downstairs...but **what** would happen to the **positive 3** exponent?

#7 $\dfrac{a^3}{b^{-4}} = a^3 b^4$

I guess it would have to change and become a **-3**...right?

Panel 4: Here too, no problem with the **x**. Leave it alone. Its exponent is **already** positive.

#8 $x^2 y^{-5} = \dfrac{x^2}{y^5}$

Just shift **y** to the basement and change the exponent.

Panel 5: #9 has got me baffled Professor Weissman! Shouldn't you move that **-5 down** and make it **+5**?

#9 $-5x^4 y^{-6} = \dfrac{-5x^4}{y^6}$

Hah! Fooled ya! **That's** no exponent! That's a **coefficient**! It stays!

Panel 6: #10 $b^{-4} c^{-3} = \dfrac{1}{b^4 c^3}$

For #10...both bases, the **b** and **c**, have negative exponents so both drop down to the denominator, the exponents change and...when nothing's left for our numerator, a **1** is understood!

Panel 7:

POWERS OF TEN
$10^{-3} =$
$10^{-2} =$
$10^{-1} =$
$10^0 = 1$
$10^1 = 10$
$10^2 = 100$
$10^3 = 1000$

OK gang! Let's use our newly acquired knowledge of negative exponents to **extend** our **Powers of Ten** table that we discussed previously.

Panel 8: Watch carefully how I simplify these.

$10^{-1} = \dfrac{1}{10^1} = \dfrac{1}{10} = .1$

$10^{-2} = \dfrac{1}{10^2} = \dfrac{1}{100} = .01$

$10^{-3} = \dfrac{1}{10^3} = \dfrac{1}{1000} = .001$

Now let's transfer the answers to our **Powers of Ten** table.

Remember patterns? What pattern do you see developing with our answers?

POWERS OF TEN

$10^{-3} = .001$
$10^{-2} = .01$
$10^{-1} = .1$
$10^{0} = 1$
$10^{1} = 10$
$10^{2} = 100$
$10^{3} = 1000$

Hmmmm...I notice something Prof. The negative exponent tells us **"how many decimal places."**

Good observation! Let me expand on that. For each answer we start with a **1** which has an **assumed** decimal on its right. Watch!

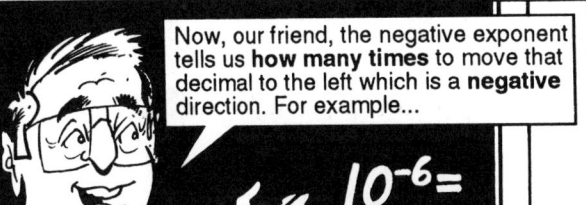

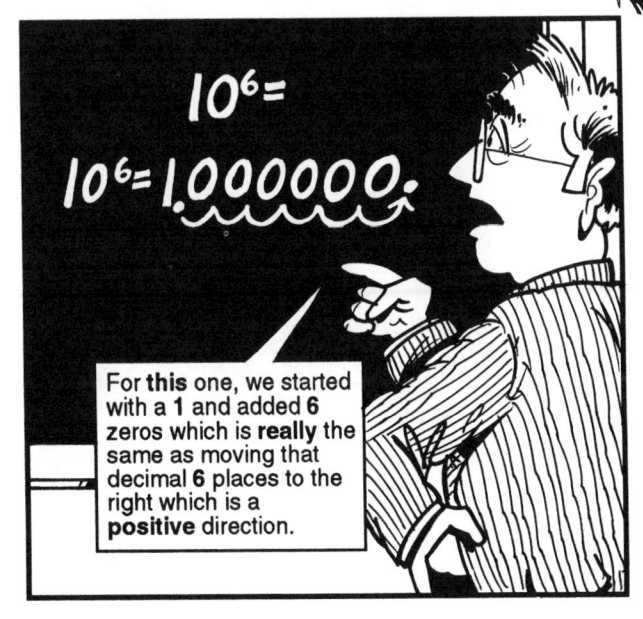

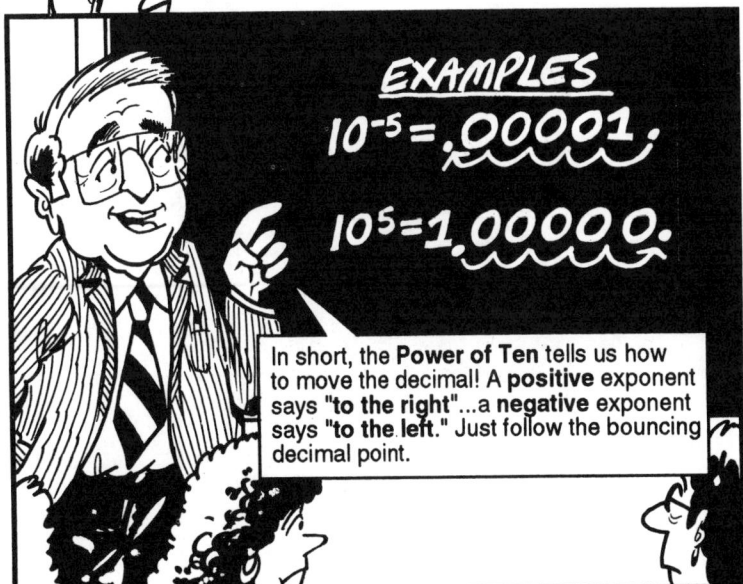

$1 = 1.$

Now, our friend, the negative exponent tells us **how many times** to move that decimal to the left which is a **negative** direction. For example...

$10^{-6} =$

...we start with a **1**...

$10^{-6} = 1.$

And insert **zeros**!

Give that man a cigar! Well...at least a candy one anyway.

$.000001$

If you recall, we **essentially** did a similar thing with **positive** Powers of Ten. Check it out.

...and move that decimal **6** places to the left and...

$10^{-6} = 1$

$10^{6} =$
$10^{6} = 1.000000.$

For **this** one, we started with a **1** and added **6** zeros which is **really** the same as moving that decimal **6** places to the right which is a **positive** direction.

EXAMPLES

$10^{-5} = .00001.$

$10^{5} = 1.00000.$

In short, the **Power of Ten** tells us how to move the decimal! A **positive** exponent says **"to the right"**...a **negative** exponent says **"to the left."** Just follow the bouncing decimal point.

OK Prof...you keep referring to these **Powers of Ten** time and time again! **What** makes them so wondrous?
Ahhhh...**That's** my cue! To help answer that question, I've brought in a **special** guest lecturer. Let's have a warm welcome for... from all the way across the hall...Science Teacher...**Professor Hilda Hooplemeister**.
Thank you...thank you! **Please** sir...a standing ovation is **completely** unnecessary!
CLAP
CLAP CLAP
CLAP
Actually...my **underwear** was kinda riding high.
HAR
HAR
HA
HA
HA
Er...uh...I see! Anyway back to these **Powers of Ten**. The good Professor asked me to stop over and help explain its significance. **Many** branches of science use either very **large** or very **small** numbers. Doing calculations with these types of numbers is **greatly** simplified by expressing these numbers in what we call **Scientific Notation**.
First, let me **define** what the term means. **Then**, the Professor will show you how we calculate with it. A number is in **Scientific Notation** when it is written as a **number between 1 and 10** (including 1) **multiplied by a Power of Ten.**
SCIENTIFIC NOTATION
Kindly take note of this chart.
MISS ALGEBRA
Roxy Reciprocal
6^2–5^2–6^2
Whoops... sorry Professor!
93,000,000 MILES
EARTH
SUN
Ahhhh...**here** it is!

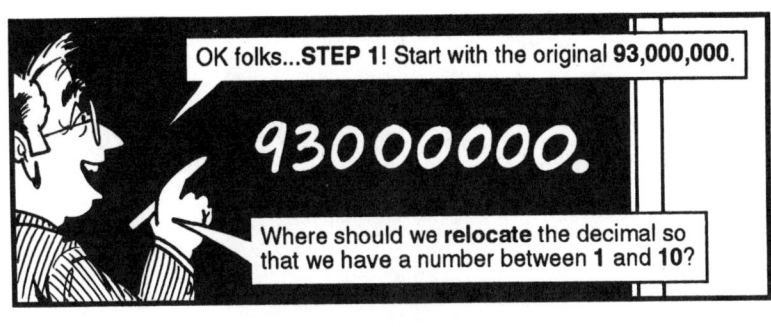

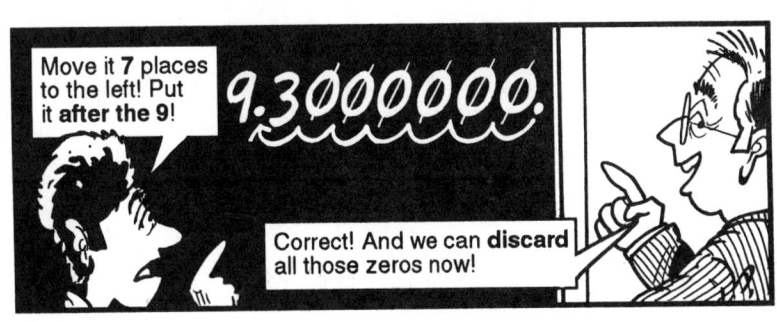

Panel 1: 93,000,000 MILES — EARTH ← → SUN

9.3×10^7

↑ A NUMBER BETWEEN 1 AND 10 ↑ MULTIPLIED BY ↑ A POWER OF 10

See how we can **rewrite** the distance in **Scientific Notation**. Let's see how we got this. Over to you Marty...er...Professor.

Panel 2: OK folks...**STEP 1**! Start with the original **93,000,000**.

93000000.

Where should we **relocate** the decimal so that we have a number between **1** and **10**?

Panel 3: Move it **7** places to the left! Put it **after the 9**!

9.3000000.

Correct! And we can **discard** all those zeros now!

Panel 4: Now, **STEP 2**! What **Power of Ten** will return that decimal to its **original** position?

9.3000000

Let's see...we moved it **7** places to the left...let's move it **back 7** places to the right. **Hey**! That's 10^7!

Panel 5: 9.3×10^7

Alright! That's how we did it! With **just** those two easy steps.

Panel 6: Now for a quick tribute to good ol' Isaac Newton and no class...**not** the guy that invented the cookies. Let's put **The Gravitational Constant** into Scientific Notation. Any volunteers for **this** way out wonder?

.0000000667 CUBIC CENTIMETERS PER GRAM-SEC^2

No problemo Prof! I watch Star Trek!

Panel 7: First, I'll move the decimal 8 places to the right. **Now** I have a number between 1 and 10...**6.67**!

.00000006.67

Panel 8: Now, we need a **Power of Ten** that will send that decimal **back** to its original position which was 8 places to the left. So **therefore**, the answer must be....

6.67×10^{-8}

Bingo! Terrific! Tell Captain Kirk you deserve a commendation. Carla, would you kindly whip out your calculator again and enter our original number?
.0000000667
Hey Professor! I can't do it! It just shows the zeros. There's no room for the 667 on the display!
Radio Shack
0.0000000
Use your EXP Key, the Exponential Entry Key and press this sequence.
6 · 6 7 EXP − 8
Radio Shack
6.67 E-8
...and I assume that everybody knows the E on calculators signifies an exponent of an assumed base of 10. In this case the exponent of 10 is -8.
My word! It works... see...
Well, if they didn't know, they know now Carla. Just remember class, most calculators won't enter more than 8 digits but once we change it to Scientific Notation, we have easy entry with the EXP Key.
Now class, note that on some calculators the Exponential Entry Key may be referred to as the EE or SCI Key! OK, Professor Hooplemeister, how about giving us one last example of Scientific Notation before you go?
With pleasure Marty... er...Professor Weissman. Class, my area of research is DNA. I've spent many long, lonely hours lately in the lab.
I've been attempting to develop techniques for distinguishing between male and female chromosomes.
That's easy Professor Hooplemeister...just pull down their genes!
HA
HA
HAR

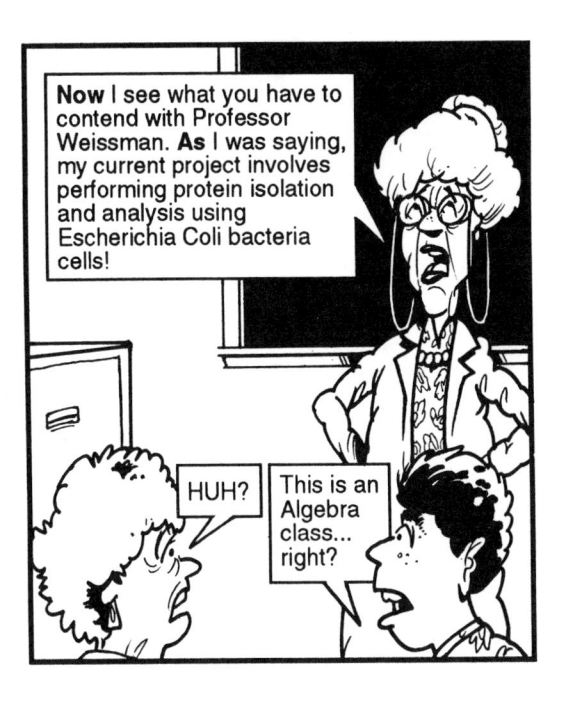

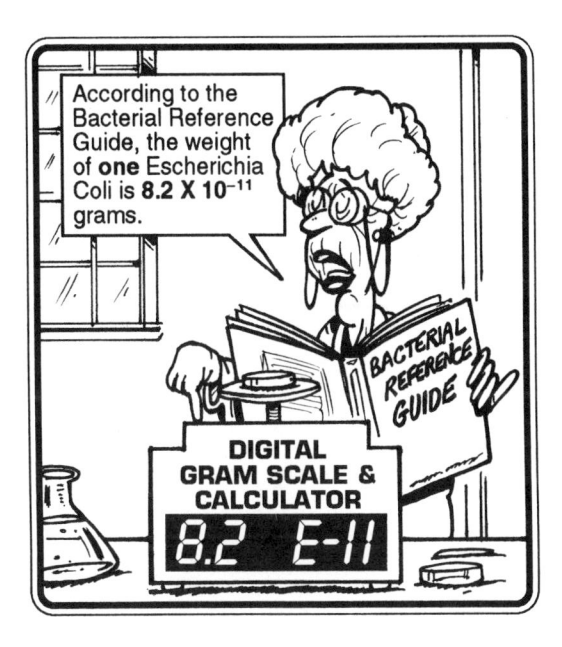

Panel 1: Now I see what you have to contend with Professor Weissman. As I was saying, my current project involves performing protein isolation and analysis using Escherichia Coli bacteria cells!

HUH? This is an Algebra class...right?

Panel 2: Last week I was in the lab as usual working on my experiment...

Hmmmm...Let's see now...sterile buffer solutions, check! Petri dishes, check! Warm nutrient agar, check! My starter kit of **300 bacteria cells**, check! I'll need at least a **million** of these babies by tomorrow. OK...into the incubator they go...check temperature... **37° Celsius** (body temperature).

Panel 3: The next day I checked my results...

Wow! Look at these colonies of bacteria cells! Let's see if we got the **one million** I needed. Hmmmm... if I **divide** the weight of these colonies of bacteria cells by the weight of **one** bacteria cell, I'll get a good idea of how **many** cells I have.

Panel 4: First...let's find the **total** weight of **all** the cells!

DIGITAL GRAM SCALE & CALCULATOR
9.1   E-5

Panel 5: Now, I **divide** the total by the weight of **one** cell.

Panel 6: According to the Bacterial Reference Guide, the weight of **one** Escherichia Coli is 8.2×10^{-11} grams.

DIGITAL GRAM SCALE & CALCULATOR
8.2   E-11

Panel 7: So the **total** amount of cells is...

Panel 8: ...hmmmm **very** interesting! Now, I'll **convert** this from **Scientific Notation**.

DIGITAL GRAM SCALE & CALCULATOR
1.1   E6

Panel 1: The positive exponent, **6**, indicates moving the decimal **6** places to the right.

$$1.1 \text{ E6} = 1.1 \times 10^6 = 1.1$$

Panel 2: Now I'll insert the zeros. **EUREKA**! I've got that **1,000,000** amount and then some!

$$1.1 \text{ E6} = 1.1 \times 10^6 = 1.100000.$$
$$= 1,100,000$$

Panel 3: Well ladies and gentlemen, that's an example of how **I** use **Scientific Notation**. Now...if you'll excuse me... I **really** must be running.

Panel 4: Thank you **so** much for stopping by Professor. If there's **anything** I could ever do...

Panel 5: Well now that you **mention** it. Dinner...my apartment... let's say about eightish! By the way...**orchids** are my favorite!

Panel 6: ...um...er...uh...er... OK folks...let's **finally** finish up those Laws of Exponents with **The Power Rule**!

LAW # 3 OF EXPONENTS
"The Power Rule"

To Find A Power Of A Power, Keep The Base & Multiply The Exponents

---

Panel 7: That's the last one! **Please** let it be the last one!

Oh it's the **last** one alright...

Panel 8: ...but it's got **3 parts**!

I knew it...I just knew it!

Panel 1: Calm down! Watch! Here we have an exponent **inside** the parentheses and **another** exponent **outside** the parentheses. The two exponents are multiplied to find our answer. Let's see why. Again, we use our oldie but goody, the **definition of exponents**.

LAW #3 OF EXPONENTS
"The Power Rule"
$$(x^a)^b = x^{a \cdot b} = x^{ab}$$
$$(x^3)^2 = x^{3 \cdot 2} = x^6$$

Panel 2: What does the exponent, **2**, tell us? That we have **two** factors of x^3.

$$(x^3)^2 = x^3 \cdot x^3 = x^{3+3} = x^6$$

Now, let's revert to **LAW #1 – The Product Rule**, which tells us to add the exponents. **Either** way, we wind up with x^6.

Panel 3: This is one variation of **LAW #3**.

$$\underbrace{(x^2 \cdot y^3)^4}_{\text{PRODUCT}} = x^8 y^{12}$$

$$\underbrace{\left(\frac{a^3}{b^4}\right)^2}_{\text{QUOTIENT}} = \frac{a^{3 \cdot 2}}{b^{4 \cdot 2}} = \frac{a^6}{b^8}$$

Here, we also use the **Power Rule** when we have a **product**. We distribute the power, that is we multiply the outside exponent by **each** of the two inside exponents...

...and we **also** use the **Power Rule** when we have a **quotient**. Again, we distribute the power.

Panel 4: But **beware**! The **Power Rule** does **not** apply with a sum!

$$(m^2 + n^3)^5 =$$
↑ ADDITION

Panel 5: OK folks...**homestretch**! Try these problems.

#1 $(a^5)^3 =$ #4 $(3mn^3)^4 =$

#2 $(b^2 c^3)^5 =$ #5 $\left(\frac{p^3}{q^2}\right)^{-1} =$

#3 $\left(\frac{d^2}{e^5}\right)^3 =$ #6 $(j^2 + k^3)^2 =$

-66-

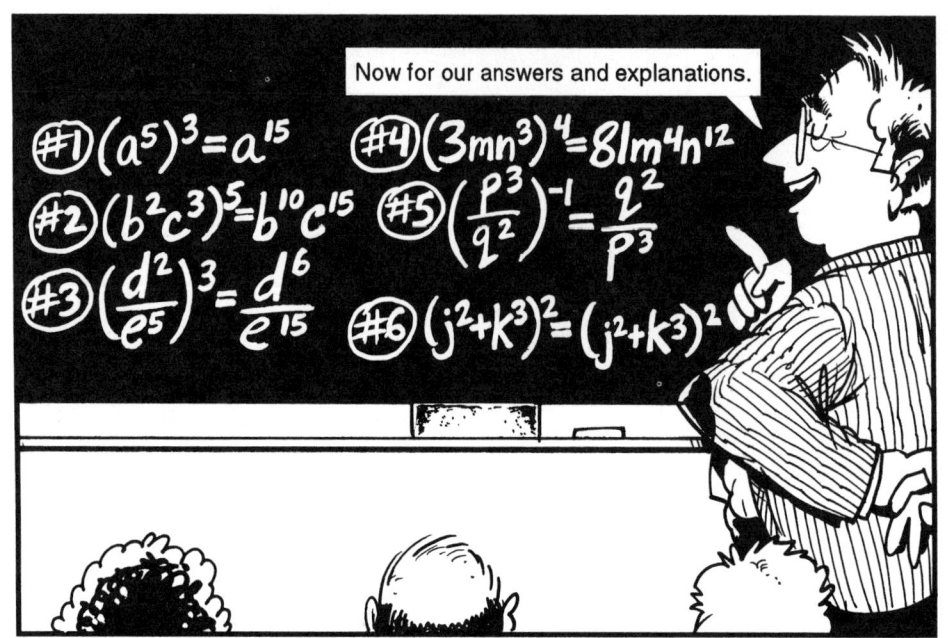

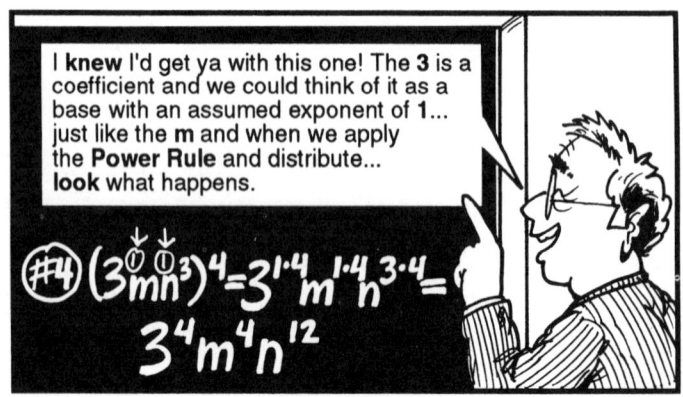

Cut the dramatics! A simple yes would have sufficed.

#5 $\left(\dfrac{p^3}{q^2}\right)^{-1} = \dfrac{p^{3\cdot -1}}{q^{2\cdot -1}} = \dfrac{p^{-3}}{q^{-2}}$

$\dfrac{p^{-3}}{q^{-2}} = \dfrac{q^2}{p^3}$

Now, for **#5**, a little tricky but no sweat. Multiply the exponents. **Now**, we pull the flapjack routine and double-flip. **Both** bases change positions and **both** exponents turn positive. Consider the negative exponent as a **spatula**. It's the device that does the flipping.

And finally, our last problem is addition. The **Power Rule doesn't** apply here.

#6 $(j^2 + k^3)^2 = (j^2 + k^3)^2$

Sit back class and take a breath! We've just covered a **lot** of ground and I'm **sure** your heads are reeling. We've touched on the three Laws of Exponents, Powers of Ten, Negative Exponents, Scientific Notation, Rounding and a whole bunch of other stuff. They include Positive and Negative Numbers, the Number Line, Ordering Signed Numbers, Absolute Values, Combining and Subtracting Signed Numbers, Multiplication and Division and on and on. **I'd** say you all got a **great** foothold in Algebra. Class, let's take a break from ingesting all this **new** material and review what we've learned. I've got a nice new batch of problems to try. Ain't I a sweetie? Good luck with them. And to all our readers especially, thanks for stopping by and I look forward to seeing **all** of you in future editions of **Professor Weissman's Laugh with Math**!

PROBLEM SET #3

See Answer Section in Back

305. The exponent tells you how many times to use the base as a ....

EXPAND THE FOLLOWING

306. 6^4

307. 8^5

308. $(-4)^6$

309. x^8

310. $(-3x)^2$

311. $-3x^2$

312. $-y^8$

313. $(ab)^2$

314. $(4x)^3$

WRITE USING EXPONENTS

315. XXXXXXXXX

316. $(-6)(-6)(-6)(-6)$

317. $-(5)(5)(5)$

318. $-7 \cdot -7 \cdot -7 \cdot -7$

319. $-7 \cdot 7 \cdot 7 \cdot 7$

320. $(5x)(5x)(5x)$

321. $5 \, x \, x \, x$

322. $(xy)(xy)(xy)(xy)(xy)$

CALCULATE EACH OF THE FOLLOWING POWERS.

323. 5^3

324. -5^2

325. 2^4

326. -2^4

327. $(-2)^4$

328. 0^{23}

329. 0^x

330. 1^{28}

331. 1^x

332. $(0.1)^3$

333. $(-0.2)^4$

334. $(1/2)^4$

335. $(-1/5)^3$

336. $(-2/3)^2$

337. $(2 \cdot 3)^2$

338. $2 \cdot 3^2$

CALCULATE EACH TERM THEN COMBINE THE RESULTS

339. $-3(-2) + 7^2$

340. $5^2 - 4(7-10)$

341. $10 - 2^4$

342. $7^2 - 4(2)(-3)$

343. $(-8)^2 - 4(1)(5)$

344. $(4-7)^2 + (-6-2)^2$

345. $5^2 - 7^2$

346. $5 - 4^3 \cdot 2$

347. $(10-3)^2 - (3-10)^2$

348. 10^5 is called a power of ......

349. With 10 as the base a positive exponent tells us "how many ...."

350. 10^9 means a 1 followed by ....

351. Any nonzero number with an exponent of zero is ....

352. The first law of exponents is called the .... rule

353. To multiply numbers with the same base, ... the base and ...... the exponents.

USE THE PRODUCT RULE, IF POSSIBLE.

354. $x^3 x^5$

355. $b^7 b^2 b^3$

356. $m^7 m$

357. $(xy)^3 (xy)^2$

358. $(4x)^5 (4x)^3$

359. $(2x)^2 (3x)^3$

360. $a^3 b^5$

361. $m^2 n^5 m^7 n$

362. $x^3 + x^4$

363. $(3e^2)(2e^3)$

364. $(-5ab^2)(3a^3b)$

365. $x^a x^b$

366. $x^{2a} x^{3a}$

367. $x^{-3} x^{-2}$

368. $x^a x^3$

369. $x^{0.5} x^{0.5}$

370. $x^{1/2} x^{1/2}$

371. $x^{-5} x^8$

372. $-2x^5(-4y^2)$

373. The second law of exponents is called the .... rule.

374. To divide two numbers with the same base .... the base and .... the exponents.

USE THE QUOTIENT RULE. IF POSSIBLE CALCULATE.

375. x^5 / x^2

376. s^7 / s

377. $6^7 / 6^2$

378. $-14b^5 / -2b^2$

379. $2^7 / 2^4$

380. $10^{10} / 10^2$

381. $a^8 b^6 / a^5 b$

382. $(4x)^8 / (4x)^2$

383. $(3x)^3 / (2x)^2$

384. $15x^8y^4z^3 / -3x^2yz$

385. $-4a^3b^7c^5 / 4ab^7c$

386. x^a / x^b

387. x^3 / x^{-5}

388. x^{-4} / x^5

389. x^{-2} / x^{-3}

ANY BASE (EXCEPT ZERO) TO THE ZERO POWER = 1. EVALUATE THE FOLLOWING.

390. x^0

391. 6^0

392. $(mn)^0$

393. $(3x)^0$

394. $(-5)^0$

395. -5^0

396. $x^0 + 5$

397. $3x^0$

398. $(3 \cdot 5)^0$

399. $3 \cdot 5^0$

WRITE WITH POSITIVE EXPONENTS. CALCULATE IF POSSIBLE.

400. 2^{-3}

401. 7^{-2}

402. x^{-9}

403. $a^{-2} b^{-4}$

404. $1 / y^{-5}$

405. $3 / m^{-2}$

406. $1 / d^{-3}e^{-1}$

407. $-2a^{-3}$

408. $(-2a)^{-3}$

409. w^{-2} / m^{-4}

410. $(3x)^{-2}$

411. The third law of exponents is called the .... rule.

412. To find a power of a power, ... the base and .... the exponents.

USE THE POWER RULE TO SIMPLIFY EACH OF THE FOLLOWING.

413. $(x^2)^3$

414. $(4^3)^5$

415. $(m)^5$

416. $(2)^5$

417. $(2m)^5$

418. $(a^4b^3)^2$

419. $\left(\dfrac{x^8}{y^2}\right)^2$

420. $(x^3 + y^2)^4$

421. $(3m^2n^3p^4)^3$

422. $(-2a^5)^3$

—70—

423. $(y^4)^{-3}$

424. $(v^{-3})^{-2}$

425. $(-abc)^4$

426. In Scientific Notation a number is written as the .... of a number between ...... and ..... times a power of ......

WRITE EACH OF THE NUMBERS IN SCIENTIFIC NOTATION.

427. 34.5

428. 123.

429. 4000

430. 0.005

431. 0.00066

432. 8.9

WRITE EACH OF THE NUMBERS AS A DECIMAL.

433. 7.04×10^4

434. 6.54×10^{-2}

435. 5.5×10^2

436. 8.8×10^{-6}

437. $1.2 \times 10^{+9}$

ROUND EACH OF THE NUMBERS TO THE NEAREST WHOLE. SOMETIMES WE SAY INSTEAD NEAREST ONE OR INTEGER.

438. 53.788

439. 138.3

440. 789.8333

441. 999.99

442. 991.119

ROUND EACH NUMBER TO THE NEAREST TEN.

443. 79

444. 379.

445. 395

446. 321.999

447. 8.888

448. 3.888

ROUND EACH NUMBER TO THE NEAREST TENTH. THE DIGIT IN THE TENTHS PLACE IS RIGHT AFTER THE DECIMAL.

449. 85.288

450. 51.754

451. 66.72

452. 43.96

MULTIPLY AND ROUND THE PRODUCT TO THE NEAREST TENTH.

453. $(-0.54)(-6.8)$

454. $(0.5)(0.5)(0.5)$

ANSWER SECTION
For Problem Sets #1, 2, & 3

1. positive

2. negative

3. integers

4. less

5. >

6. right

7. left

8. > because on the number line, +9 is more to the right.

9. > because +6 is more to the right on the number line.

10. > 0 is more to the right.

11. > +7 is more to the right.

12. < On the number line -3 is to the left of 0 and so -3 is less than 0.

13. < On the number line -4 is to the left of -1 thus -4 is less than -1.

14. absolute value

15. positive / zero

16. 9

17. 9

18. 0

19. sign (direction) / magnitude (absolute value).

20. positive (+)

21. add / absolute values
Use this rule for the next 4 problems (22. - 25.)

22. +21

23. -17

24. -6

25. -10

26. subtract / larger
Use this rule for the problems that follow.

27. -4

28. -8

29. +8-15=-7

30. -3
#1 Combine the negatives: -1-3-5=-9
#2 Combine the positives: +2+4=+6
#3 Combine those results: -9+6=-3

31. +3
#1 Sum the negatives: -2-4 = -6
#2 Sum the positives: +1+3+5 = +9
#3 Sum the results: -6+9= +3

32. 0
    These numbers -10 and +10 are opposites for addition (combining).

33. -3.6
    Subtract absolutes then use sign of the larger (-)

34. +5/24
    #1 Use LCD=24   -16/24 +21/24
    #2 Now combine   +5/24

35. +10 , because -10 +10 =0

36. zero (0)

37. sign
    Zero is neither positive nor negative.

38. change
    Use this rule for the next 6 problems (39. - 44.)

39. +7

40. -8

41. -1
    Method #1: Change BOTH signs inside parentheses, then combine. +7 -8 =-1

    Method #2: Combine within parentheses first then change the sign. -(+1) = -1

42. +10
    Method #1
    change both signs   4+6 = +10
    Method #2
    combine inside 1st -(-10)= +10

43. -13
    #1 change both signs: -3-10
    #2 combine: -13

44. +2
    #1 change all 3 signs: +1-2+3
    #2 combine: +4-2=+2

45. change

46. -7

47. -3

48. -2 -5 = -7

49. -6+9 = +3

50. -1+2-3 = -4+2 = -2

51. terms

52. -21
    -6 -7 -8 = -21

53. +21
    +10 +4 +7 = +21

54. +36
    +40 -8 +10 -6
    +40 +10 -8 -6
    +50 -14 = +36

-73-

55. 48
High - Low = 88 - 40 = 48

56. 38
33 - (-5) = 33 + 5 = 38

57. 5/7
2/7 + 3/7 = 5/7

58. -1/5
+2/5 -3/5 = -1/5

59. +2.1
-7.9 +10
+10.0 -7.9 = +2.1

60. +21.6
+20 +1.6
+20.0 +1.6 = +21.6

61. +17/16
+7/8 +3/16   LCD=16
+14/16 +3/16
+17/16

62. -0.001
+0.999 -1
-1.000 +0.999
-0.001

63. +15
+46-16+5-20
+51-36 = +15

64. 880
+850 -(-30)
+850 +30 = 880

65. 14777
+14495 - (-282)
14495 + 282
14777

66. 29
88 - 59 = 29

67. +18
+6 - (-12) = +6 +12 = +18

68. 1008
-282-(-1290)
-282+1290=1008

69. letters (literal factors) / exponents.

70. 6xy, -5xy

71. 6ab, -ab, 4ba
NOTE: The order of the letters is not important. 4ba is the same as 4ab.

72. $5m^2$, $3nm^2$

73. numerical coefficient

74. the letters a and b

75. 1 or +1

76. -8

77. +1

78. -1

79. numerical coefficients / literal factors (letters)

80. Unlike

81. combined (added/subtracted)

82. 7yx

83. -1

84. literal factors

85. exponents

86. -11x

87. -10abc

88. +1xy -8xy = -7xy

89. +3a^2

90. +2a -7a -5b -1b
    -5a -6b

    NOTE: Since the order of terms is NOT important, another answer is -6b -5a

91. +5mn -8mn = -3mn

    NOTE: Since the order of factors is not important, another answer is -3nm

92. +1a^2 +1a^2 = +2a^2

93. The student tried to combine unlike terms.

94. He added the exponents.
    +5b^3 +7b^3 = +12b^3

95. They are UNLIKE terms and they cannot be combined.

96. 5

    #1 Combine within absolutes | -5 |
    #2 Write the absolute value: 5

97. 9

    #1 Simplify each absolute 7 + 2
    #2 Combine   9

98. -7
    #1 The absolute value of -7 is 7
    #2 The negative at the start of the problem makes the final answer -7

99. -7
    This is similar to the previous problem. The absolute is positive but the negative at the start of the problem makes the answer negative.

100. -5
    Again, simplify the absolute then make the answer negative.

    - | -7 +2 | = - | -5 | = -5

101. -12

#1 combine the absolutes 12
#2 Use the common sign -12

102. +3

-3-7-2+5+1+9
-12 +15 = +3

103. 0

+8+1-3-2-4
+9 -9 = 0

104. +8d
#1 Combine the coefficients
#2 Keep the common factor d

105. +13m
+5m +7m +1m = +13m

106. -13m
-5m -7m -1m = -13m

107. -3n

108. 10m +3n
#1 Combine the m terms
#2 Combine the n terms

NOTE: An alternate answer is
3n +10m

WARNING: Do not try to simplify any further. 3n and 10m are UNLIKE terms.

109. -4m -3n
Same warning as before. STOP!!

110. The terms cannot be combined

111. $10c^2$
NOTE: the answer 'looks like' the two terms that were combined ; it has the same literal factors.

112. +11mn

113. 19ab
The missing coefficient is -1

114. +11cd
These ARE like terms.

115. -5a -12b +5ab

116. $-3xy^2$

117. The terms are UNLIKE. They cannot be combined

118. 0
These terms are opposites for addition.

119. -4x
+5x -9x -7y +2y +5y = -4x
NOTE: the y terms cancel out.

120. -4 -5x
5 -9 -3x -2x = -4 -5x
Alternate answer: -5x -4

121. 0
10m -10m -3 -2 +5
0+0=0
Everything cancels out !

122. +5 or 5

123. -6

124. -1
Method #1 Combine within parentheses then change the sign. -(+1) = -1
Method #2 Change both signs inside parentheses then combine.   +5 -6 = -1

125. +13
Remember the poem : To subtract, change the sign and combine.   10 +3 = 13

126. -4
When a + appears outside (on the left of) parentheses, just remove the parentheses. DO NOT CHANGE any signs inside.

127. +8

128. -1
-6 +5 = -1

129. 16
+12 -(-4)
+12 +4 =16

130. -4
There are three terms in this problem. Simplify each term then COMBINE the results.
+4 -3 -5
+4 -8 = -4

131. +46
+40 -(-5) +1
+40 +5 +1 = +46

132. +5a
Change the sign.

133. -2b +1
Change BOTH signs.

134. +10e
Method #1 +3e +7e =+10e
Method #2   -(-10e) =+10e

135. +7x
+2x +5x = +7x

136. -4a -12
+2a -7 -5 -6a
+2a -6a -7 -5 = -4a -12

137. 5m-6
+3m -7 +2m +1
+3m +2m -7 +1 = 5m -6

138. 2a -4b
NOTE: The parentheses are said to be 'not essential.' This means that you can remove them without affecting the answer. We can just 'erase' them.

-77-

139. $8a - 11b$
$2a - 4b + 6a - 7b = 8a - 11b$

140. $-4a + 3b$
$+2a - 4b - 6a + 7b = -4a + 3b$

141. $10x$
$7x - (-3x)$
$7x + 3x = 10x$
NOTE: When a problem says subtract 'from' be sure to write the 'from' expression first.

142. $5a - 7b$
$(7a - 2b) - (2a + 5b)$
$7a - 2b - 2a - 5b = 5a - 7b$

143. $-5x$
$(-3x) + (-2x)$
$-3x - 2x = -5x$

144. $-1x$
$(-3x) - (-2x)$
$-3x + 2x = -1x$

145. $7.46x$
$2.30x + 5.00x + 0.16x = 7.46x$
Write each term with the same amount of decimals. Remember that 5 can be written as 5.00

146. $36.6x$
$40.0x - 3.4x = 36.6x$
Same comment as in the previous problem.

147. $21.29y$
$23.00y - 1.71y = 21.29y$

148. $(22/15)x$
Use the LCD 15 and change both fractions then combine.
$(10/15)y + (12/15)y = (22/15)y$
NOTE: The answer could be changed to a mixed number. Generally improper fractions are preferred in Algebra.

149. $(45/8)y$
#1 Change coefficients to fractions
#2 Use LCD to combine

$(5/2)y + (25/8)$
$(20/8)y + (25/8)y = (45/8)y$

150. NOTE: To combine either
#1 change 3/4 to a decimal or
#2 change 2.5 to a fraction.

#1 $.75a + 2.5a = 3.25a$

#2 $(3/4)a + (5/2)a$
$(3/4)a + (10/4)a$
$(13/4)a = (3\ 1/4)a$

151. $-4p$
Combine the numerical coefficients and keep the common literal factor p.

152. $+4p$
Same comment as in the previous problem.

-78-

153. -7, -3, 0, +2, +5
The numbers are put in order as they appear on the number line.

154. +4, +5, +9
#1 Simplify each expression
+5, +9, +4
#2 Put results in order
+4, +5, +9

155. -6.5, -6.1, 0, +7.01
#1 Simplify each expression
-6.5, 0, +7.01, -6.1
#2 Put results in order
-6.5, -6.1, 0, +7.01

156. -12
-5 -7 = -12

157. 12
5 + 7 = 12

158. -1
-4 +3 = -1

159. 1
4 - 3 = 1

160. -3
#1 Remove the brackets
#2 Remove the parentheses
- (- [-3]) = - (+3) = -3

161. -5
- (- [-5]) = - (+5) = -5

162. -4a+9b-3c

(3a+8b-5c) + (-7a+b+2c)
3a+8b-5c-7a+1b+2c
4a+9b-3c

163. 5x-10y

(3x-2y) + (6x-7y) + (-4x-y)
3x-2y+6x-7y-4x-y
5x-10y

164. $4x^2$ -10x

$(3-7x-x^2) + (5x^2-3x-3)$
+3 -7x $-x^2$ $+5x^2$ -3x -3
$4x^2$ -10x
NOTE: The constants -3 and +3 cancel.

165. 25 - 12 = 13

166. 3x+3y

(5x+4y) - (2x+y)
5x+4y -2x-y
3x+3y

167. -10
+10 +(-10)= +10-10 =0

168. -4x+7y
Add the opposites.

169. 30
Note how it is changed into a subtraction problem.
50 - 20 = 30

-79-

170. 24
   +10 - (-14)
   +10+14 = 24

171. -15x
   -7x -(8x)
   -7x-8x = -15x

172. 13
   30 - 17 = 13

173. -25
   -40 - (-15)
   -40+15 = -25

174. 14a
   7a - (-7a)
   7a +7a = 14a

175. -4x -7y

   3x-5y - (7x+2y)
   +3x-5y-7x-2y
   -4x -7y

176. -5a -6b

   2a-5b +(-7a-b)
   +2a-5b-7a-1b
   -5a-6b

177. X can be either -5 or +5

   Replace X with -5
   |-5|=5

   Replace X with +5
   |+5|=5

178. X = 7
   If you replace x with 7 you get
   |7 - 7| which = |0| = 0

179. YES.

   Replace x with -8
   Is |-8| < 9 ?
   8 < 9   YES !

180. YES.

   Replace x with -7
   Is |-7| > 4 ?
   7 > 4   YES !

181. NO.

   Replace x with -5
   Is |-5-1| < 6 ?
   |-6| < 6
   6 < 6 No. 6=6 !

182. YES.

   Replace x with +1
   Is |2 - (-3)| > 0 ?
   |+2 +3| > 0
   |+5| > 0
   5 > 0 YES !

For the next 3 problems, recall that the perimeter of a rectangle is twice the length + twice the width. The formula is written: P = 2L + 2W

183. 24

$P = 2(5) + 2(7)$
$P = 10 + 14$
$P = 24$

184. 6X

$P = 2(X) + 2(2x)$
$P = 2X + 4X$
$P = 6X$

185. $P = 2X + 2Y$

186. 4X-12

The perimeter of the triangle is the sum of all three sides.
$(2x-5) + (x) + (x-7)$
$2x-5+x+x-7$
$4x-12$

187. 85

Multiply the amount of nickels by 5 since each is worth 5 cents.
$5(17)=85$

188. 5X

Again, each nickels is worth 5 cents.
$5(X) = 5X$

189. 15X

$5(3X)=15X$

190. 70X

Multiply the amount of dimes by 10 since each dime is worth 10 cents.
$10(7x)=70X$

191. 39
$22+17=39$

192. 2X-3

$X+(X-3)$
$X+X-3$
$2X-3$

193. 20
$49-29 = 20$

194. X-29

195. NO.
Replace X with 7.
$7 < 5$ is false.

196. YES.
Replace X with -4.
$-4 > -7$ is true.

197. YES.
Replace X with -10.
$-10 < 0$ is true.

198. YES.
Replace X with -3.
$0 > -3$ is true

199. YES.

Replace X with -3.
-7 < -3 is true.

200. 0 > -7

201. 7 > -10

202. X < 0

This says "X is less than 0" and, in fact is how we write "X is NEGATIVE."

If we want to say X is POSITIVE we write X > 0, which is read "X is greater than 0."

203. X > -5

204. YES.

6 is between 5 and 12.

205. YES.

-3 is between -7 and 0.

206. NO.

-8 is not between -5 and -1

207. NO.

Replace X with -2.
-3 < X - 6 < +3
-3 < -2 -6 < +3
-3 < -8 < +3   False.
-8 is NOT between -3 and +3

208. NO.

Replace X with -3
-8 < X + 5 < 0
-8 < -3 +5 < 0
-8 < +2 < 0 False.
+2 is NOT between -8 and 0

209. YES.

-7-7 $\geq$ -14
-14 $\geq$ -14   True
The symbol $\geq$ means GREATER THAN OR EQUAL TO. -14 is not GREATER THAN but it is EQUAL TO -14.

210. NO.

8 - (-8) $\leq$ 0
8 + 8 $\leq$ 0
16 $\leq$ 0   False

211. YES.

-6 +4 $\geq$ -2
-2 $\geq$ -2   True

212. YES.

-7 $\leq$ -7 < +7   True
-7 is EQUAL TO -7

213. NO.

-7 $\leq$ +7 < +7   False
+7 is NOT less than +7

-82-

214. Product

215. Factors

216. +56

217. +30

218. -80

219. 0

220. -9

221. +24

222. -6

223. +24

224. -1

225. -168

226. +24

227. -120

228. Quotient / divisor

229. +3

230. +1/3

231. +3

232. -3

233. +1/3

234. 0

235. 0

236. undefined

237. undefined

238. -1

239. +4/5

240. -40

241. +1

   #1. Simplify numerator -6
   #2. Simplify denominator -6
   #3. Divide -6/-6 = +1

242. -4/3

   #1. Numerator = +8
   #2. Denominator = -6
   #3. Divide : +8/-6
   #4. Reduce: -4/3

243. -4/9

   #1 Numerator = 12
   #2 Denominator = -27
   #3 Divide: 12/-27
   #4 Reduce: -4/9

244. ZERO

   Zero may appear in the numerator but not in the denominator.

245. X cannot be 0.

If X would be zero then the denominator would be zero.

246. X cannot be 4.

When X=4 the denominator becomes 4-4 or ZERO.

247. X cannot be -5.

When x=-5 the denominator becomes -5+5 or ZERO.

248. -4
$(-3)(-4) = 12$

249. -2
$(+15)(-2) = -30$

250. 0
$(0)(-8) = 0$

251. -9
$(-9)(-9) = +81$

252. -5
-10 / -5 = +2

253. -10
-10 / -10 = +1

254. +10
-10 / +10 = -1

255. -50
-10 / -50 = +1/5

256. -15

-3 +2(-6)
-3 -12 = -15

257. -12

-3(+2) -6
-6 -6 = -12

258. -10

259. +14

-1(-2) -3(-4)
+2   +12 = +14

260. +1

-1 -2(-3) -4
-1 +6 -4
-5 +6 = +1

261. +9

-1 -2 -3(-4)
-1 -2 +12
-3 +12 = +9

262. -10

(-1)(-2)(-3) -4
-6 -4 = -10

263. +13.11

264. -0.008

-84-

265. +1/6

Recall from Arithmetic that you multiply 'across.' That is, multiply the numerators then multiply the denominators.

266. +7

Did you remember that when you multiply by 10 all you need do is move the decimal one place TO THE RIGHT ?

267. -356

When you multiply by 100 you need only move the decimal two places to the right.

268. +1/9

#1 Simplify the numerator:
 -4+3 = -1
#2 Simplify the denominator:
 -2-7 = -9
#3 Divide: -1/-9 = +1/9

269. -1

#1 Simplify the numerator:
 +9+1 = +10
#2 Simplify the denominator:
 -1-9 = -10
#3 Divide: +10/-10 = -1

270. -2

#1 Numerator: -1-1 = -2
#2 Denominator: (-1)(-1)=+1
#3 Divide: -2/+1 = -2

271. + (POSITIVE)
There are 2 negative factors an EVEN amount.

272. - (NEGATIVE)
There are 3 negative factors an ODD amount.

273. - (NEGATIVE)
There is only one negative factor an ODD amount.

274. 2x + 2y

275. -3a -3b

276. 14x - 7

277. -15x - 10

278. 4a +4b -4c

279. -24x + 6

280. 6xy -2x

281. -15a + 6ab

282. 10a -15b +20c

283. -6x +10y -2z

284. -2x -8
10x -5 -12x -3
10x -12x -5 -3 = -2x -8

285. $18x + 3$

$15x + 10 - 7 + 3x$
$18x + 3$

286. $+2b$

$4a - 2b - 4a + 4b$
$4a - 4a - 2b + 4b$
$+2b$

287. $-10a - 10b$

$6ab - 10a - 6ab - 10b$
$6ab - 6ab - 10a - 10b$
$-10a - 10b$

288. $30 - 14x - 7xy$

$30 - 15x - 7xy + 1x$
$30 - 15x + 1x - 7xy$
$30 - 14x - 7xy$

NOTE: The answer has 3 terms and can be written many ways. For example: $-14x - 7xy + 30$

289. $-17x + 9y$

Note: The FROM expression is written first.
$5(y-x) - 4(3x-y)$
$5y - 5x - 12x + 4y$
$-5x - 12x + 5y + 4y$
$-17x + 9y$

290. $2x + 21$

NOTE: Subtract in the given order.
$5x - 3(x-7)$
$5x - 3x + 21$
$2x - 21$

291. $-4x + 28$

NOTE: Again, write the expressions in the order given.
$0 - 4(x-7)$
$0 - 4x + 28$
$-4x + 28$

292. $2x - 7y - 35$

$2(x-7y) + 7(y-5)$
$2x - 14y + 7y - 35$
$2x - 7y - 35$

293. $4x - 10$

NOTE: Perimter= 2L + 2W
$2(x) + 2(x-5)$
$2x + 2x - 10$
$4x - 10$

294. $25x - 30$

$5(x) + 10(2x-3)$
$5x + 20x - 30$
$25x - 30$

295. 350 miles

$R \times T = D$
$50(7) = 350$

296. 7x -35

    R X T = D
    (x-5)7 = 7x -35

297. $48.00

    R X P = Interest
    .08(600)
    48.00

298. .04x

    .04(x)= .04x

299. 12 - .04x

    .04(300-x)
    12.00 - .04x
    12-.04x

300. .12x - 18

    .06(2x-300)
    .12x -18.00
    .12x -18

301. .25x -24

    #1 Find 7% of x.
    .07 (x)= .07x
    #2 Find 6% of 3x-400.
    .06(3x-400)
    #3 Sum the results.
    .07x + .06(3x-400)
    #4 Simplify
    .07x +.18x -24.00
    .25x -24

In each of the next 3 problems we multiply the amount of tickets by their respective prices.

302. $116

    $5(12) + $7(8)
    60 + 56 = 116

303. 5x +7y

    #1 Multiply the x tickets by 5
    #2 Multiply the y tickets by 7
    #3 Combine

304. 19x - 7

    #1 Multiply x tickets by 5
    5(x) =5x
    #2 Multiply 2x-1 tickets by 7
    7(2x-1)=14x-7
    #3 Combine
    5x + 14x -7
    19x -7

305. factor

306. 6·6·6·6

307. 8·8·8·8·8

308. (-4)(-4)(-4)(-4)(-4)(-4)

309. X X X X X X X

310. (-3x)(-3x)

311. $-3xx$

NOTE: Only the x is written twice. The -3 is NOT part of the base.

312. $-yyyyyyy$

NOTE: The negative is NOT part of the base.

313. $(ab)(ab)$

314. $(4x)(4x)(4x)$

315. x^{10}

316. $(-6)^4$

317. -5^3 Without parentheses!

318. $(-7)^4$

319. -7^4

320. $(5x)^3$

321. $5x^3$

322. $(xy)^5$

323. $(5)(5)(5) = 125$

324. $-(5)(5) = -25$

NOTE the one negative and 2 factors of 5.

If your answer was +25 you probably thought (-5)(-5). But that would have been written differently as $(-5)^2$ with parentheses around the base -5.

325. 16

$(2)(2)(2)(2) = 16$

326. -16

$-(2)(2)(2)(2) = -16$

This problem is similar to problem 324. One negative and four twos.

327. +16

$(-2)(-2)(-2)(-2) = +16$

328. 0

NOTE: Zero to any exponent (except 0) = 0

329. 0

See problem 328.

330. 1

NOTE: One to any exponent = 1.

331. 1

Again, one to any exponent = 1.

332. 0.001

$(0.1)(0.1)(0.1) = 0.001$

333. +0.0016

334. 1/16

335. -1/125

336. +4/9

337. 36

   #1 Simplify within parentheses.
   (6)²
   #2 Do the exponent.
   36

338. 18

   #1 Do the exponent.
   2 · 9
   #2 Multiply.
   18

339. 55

   #1 Simplify each term.
   +6 +49
   #2 Combine
   +55

340. 37

   +25 -4(-3)
   +25 +12 = 37

341. -6

   10 - 16 = -6

342. 73

   +49 +24 = +73

343. 44

   +64 -20 = +44

344. 73

   (-3)² + (-8)²
   +9 + 64 = 73

345. -24

   (5)(5) - (7)(7)
   +25 - 49
   -24

346. -123

   5 - (4)(4)(4)(2)
   +5 -128
   -123

347. 0

   (7)² - (-7)²
   49 - 49
   0

348. Ten

349. zeros follow a one

350. 9 zeros

351. 1

352. product

-89-

353. keep / add

354. $x^{3+5} = x^8$

355. $b^{7+2+3} = b^{12}$

356. $m^7 m^1 = m^{7+1} = m^8$

357. $(xy)^5$

358. $(4x)^8$

359. $(2x)^5$

360. $a^3 b^5$
The bases are DIFFERENT. The product rule does NOT apply.

361. $m^{2+7} n^{5+1} = m^9 n^6$

362. $x^3 + x^4$
This is not a multiplication problem. The product rule does NOT apply.

363. $6e^5$
Multiply the coefficients, add the exponents.

364. $-15 a^4 b^3$

365. x^{a+b}

366. $x^{2a+3a} = x^{5a}$

367. $x^{-3-2} = x^{-5}$

368. x^{a+3}

369. x

$x^{0.5+0.5} = x^{1.0} = x$

370. x

$x^{1/2+1/2} = x^1 = x$

371. $x^{-5+8} = x^{+3}$

372. $+8 x^5 y^2$

373. Quotient rule

374. keep / subtract

375. $x^{5-2} = x^3$

376. $s^{7-1} = s^6$

377. $6^{7-2} = 6^5$
Keep the base 6.

378. $+7 b^3$

379. 8

$2^{7-4} = 2^3 = (2)(2)(2) = 8$

380. $10^{10-2} = 10^8 = 100000000$

381. $a^3 b^5$

382. $(4x)^{8-2} = (4x)^6$

383. The Quotient rule does not apply because the bases are different. However, it IS possible to do it another way.

#1 Simplify the numerator:
$(3x)(3x)(3x) = 27x^3$
#2 Simplify the denominator:
$(2x)(2x) = 4x^2$
#3 Now apply the Quotient rule:
$27x^3 / 4x^2 = 27x / 4$

384. $-5x^6y^3z^2$

385. $-a^2c^4$

$-1a^{3-1}b^{7-7}c^{5-1}$
$-1a^2b^0c^4$
$-a^2c^4$
NOTE: $b^0 = 1$

386. x^{a-b}

387. x^8

$x^{3-(-5)}$
$x^{3+5} = x^8$

388. x^{-9}

$x^{-4-5} = x^{-9}$

389. x

$x^{-2-(-3)}$
$x^{-2+3} = x^1 = x$

390. 1

391. 1

392. 1

393. 1

394. 1

395. -1
The negative is NOT part of the base as it is in the previous problem. Thus, $5^0 = 1$ and the negative makes the answer -1.

396. 6
$1 + 5 = 6$

397. 3
$3(1) = 3$
The 3 is NOT part of the base as it is in problem 393.

398. 1
$(15)^0 = 1$

399. 3
$3(1) = 3$

400. $1 / 2^3 = 1 / 8$

401. $1 / 7^2 = 1 / 49$

402. $1 / x^9$

403. $1 / a^2b^4$

404. y^5

405. $3m^2$

406. $d^3 e^1 = d^3 e$

407. $-2/a^3$

The negative 2 stays in the numerator. It is NOT an exponent.

408. $1/(-2a)^3$ or $-1/8a^3$

The denominator can be simplified. $(-2a)(-2a)(-2a) = -8a^3$ giving us a final answer of $-1/8a^3$

Note: When there is one negative we usually put it at the beginning of the answer.

409. m^4/w^2

410. $1/(3x)^2$ or $1/9x^2$

Here too, the denominator can be simplified. $(3x)(3x) = 9x^2$ giving us a final answer of $1/9x^2$

411. Power

412. keep / multiply

413. $x^{2(3)} = x^6$

414. $4^{3(5)} = 4^{15}$

415. $m^{1(5)} = m^5$

416. $2^{1(5)} = 2^5 = 32$

417. $2^5 m^5 = 32 m^5$

418. $a^8 b^6$

419. x^{16}/y^4

$x^{8(2)} / y^{2(2)} = x^{16}/y^4$

420. $(x^3 + y^2)^4$

This expression cannot be simplified using the power rule because the power rule does not apply to a sum.

421. $27 m^6 n^9 p^{12}$

Note the coefficient 3 is also raised to the third power: $3^3 = 27$

422. $-8 a^{15}$

$(-2)^3 a^{5(3)} = -8 a^{15}$

423. y^{-12}

$y^{4(-3)} = y^{-12}$

—92—

424. v^{+6}

$v^{-3(-2)} = v^{+6}$

425. $a^4b^4c^4$

$(-1abc)^4$
$(-1)^4 a^4 b^4 c^4$
$+1\, a^4 b^4 c^4$

426. product / one / ten / ten

427. $3.45 \times 10^{+1}$

Move the decimal 1 place to the left then compensate by multiplying with 10^{+1}

428. $1.23 \times 10^{+2}$

Move the decimal to the left twice and compensate with $\times 10^{+2}$

429. 4.000×10^3

430. 5.0×10^{-3}

Move the decimal to the right 3 places and compensate using 10^{-3}

431. 6.6×10^{-4}

432. 8.9×10^0

The number 8.9 is already between 1 and 10. The decimal need not be moved. That's why we use 10 with a ZERO exponent.

433. 70400.

The exponent 4 says "move the decimal 4 places to the right." We add 2 zeros.

434. 0.0654

The exponent -2 says "move the decimal 2 places to the left."

435. 550.

436. 0.0000088

437. 1200000000

438. 54

The rounding digit is 3. Since the next digit (decision digit) is 7 we add 1 to the rounding digit.

439. 138

The decision digit is 3 so we do NOT add one to the rounding digit.

440. 790

The decision digit is 8 which means that we add 1 to the rounding digit 9.

But 789 + 1 = 790.

441. 1000

Add 1 to 999. 999+1=1000.

-93-

442. 991

The decision digit is 1 so the rounding digit stays 1.

443. 80   79
         +1
         80

Add 1 to the rounding digit 7. Note that the decision digit becomes zero.

444. 380   379
           +1
           380

445. 400   395
           +1
           400

The decision digit again becomes zero.

446. 320

The rounding digit stays and the decision digit becomes zero and all digits after the decimal are dropped!

447. 10   08.8888
          +1
          10

The rounding digit increases by 1, the decision digit becomes zero and all digits after the decimal are dropped.

448. 0

Drop the decision digit and all digits after the decimal.

449. 85.3   85.288
            +1
            85.3

Plus 1 to the rounding digit and drop all those digits after it.

450. 51.8   51.754
            +1
            52.8

451. 66.7

452. 44.0   43.96
            +1
            44.0

Do not write the answer as 44. The .0 is necessary. When you round to the tenths place, you need a digit in that place in the answer.

453. 3.7
3.672 = 3.7 rounded (tenth)

454. 0.1
0.125 = 0.1 rounded (tenth)